MW01702041

TELECOURSE GUIDE TO ACCOMPANY

FOR ALL PRACTICAL PURPOSES

Introduction to Contemporary Mathematics

THIRD EDITION

Consortium for Mathematics and Its Applications
(COMAP)

W. H. FREEMAN AND COMPANY

NEW YORK

Front Cover: Wassily Kandinsky, *Swinging*, 1925. Tate Gallery, London/Art Resource, NY
(Copyright 1993 ARS, NY.)

Major funding for the FOR ALL PRACTICAL PURPOSES: INTRODUCTION TO CONTEMPORARY MATHEMATICS telecourse and for the television series, FOR ALL PRACTICAL PURPOSES, has been provided by the Annenberg/CPB Project. Additional funding has been provided by the Carnegie Corporation of New York. The Alfred P. Sloan Foundation funded the development of the textbook.

Printed in the United States of America

ISBN 0-7167-2530-4

1 2 3 4 5 6 7 8 9 0 MB 9 9 8 7 6 5 4

Contents

Preface

For All Practical Purposes: Introduction to Contemporary Mathematics is an introductory mathematics course for students in the liberal arts or other nontechnical curricula. The course consists of twenty-six half-hour television shows, the textbook, and this *Telecourse Guide*. This series shows mathematics at work in today's world. Part of the power of television is its ability to bring you to locations where you can see mathematics at work solving practical problems.

For All Practical Purposes aims to develop conceptual understanding of the tools and language of mathematics and the ability to reason using them. We expect most students will have completed elementary algebra and some geometry in high school. We do not assume students will be pursuing additional courses in mathematics, at least none beyond the introductory level.

The organization of this *Telecourse Guide* parallels the television shows. For each of the five subject-area clusters, the television series contains an overview show and four additional programs. For each of the programs in the series, you will find a section in this guide.

How to Use This Guide

For each program in the series, this Guide includes

- an outline and summary of the program.

- a list of learning objectives.

- sample examinations with answers. The examinations consist of multiple-choice questions and long-answer questions.

Following each section in the textbook is a list of review vocabulary. This list of terms will help you understand the most important concepts. Before viewing a program, you will find it valuable to read over the respective section in this *Telecourse Guide*, including the list of objectives. The objectives will provide you with an idea of the topics to be covered in each program.

The chart on page vii lists the appropriate textbook chapter to be read in conjunction with each show. Textbook page numbers are included for easy reference. You will find additional explanation of key concepts in the summary sections of the *Telecourse Guide*, and additional worked-out examples in those same sections.

The same mathematical concepts are explained and illustrated in slightly different ways in the television programs, *Telecourse Guide*, and textbook. Each part of the presentation of a subject area complements the others. For example, you will find that the worked-out examples and discussions contained in this *Telecourse Guide* may give different emphases for a particular topic than does the textbook discussion. Thus, to get the greatest benefit from this course, you should study all parts of the instructional package.

This *Telecourse Guide* and the television series include a section on Computer Science that is no longer covered in the third edition of *For All Practical Purposes*. The Computer Science television episodes and exercises in this *Telecourse Guide* may, however, be useful as review materials for the Coding Information chapters.

Reference Chart to Textbook

Television Program	Telecourse Guide Pages	Textbook Chapter	Textbook Pages
Overview Show • Management Science	1–2	Part I Introduction	1–4
Show One • Street Smarts	3–10	1 Street Networks	5–32
Show Two • Trains, Planes, and Critical Paths	11–20	2 Visiting Vertices	33–74
Show Three • Juggling Machines	21–32	3 Planning and Scheduling	75–106
Show Four • Juicy Problems	33–41	4 Linear Programming	107–141
Overview Show • Statistics	43	Part II Introduction	143–144
Show One • Behind the Headlines	44–50	5 Producing Data	145–178
Show Two • Picture This	51–56	6 Describing Data	179–214
Show Three • Place Your Bets	57–61	7 Probability	215–244
Show Four • Confident Conclusions	62–68	8 Statistical Inference	245–272
Overview Show • Social Choice	69–70	Part IV Introduction	329–332
Show One • The Impossible Dream	71–75	11 Social Choice	333–358
Show Two • More Equal Than Others	76–78	12 Weighted Voting Systems	359–398
		13 Fair Division	399–426
		14 Apportionment	427–460
Show Three • Zero-Sum Games	79–82	15 Game Theory	461–490
Show Four • Prisoners' Dilemma	83–86	15 Game Theory	461–490
Overview Show • On Size and Shape	87–88	Part V Introduction	491–492
Show One • How Big Is Too Big	89–93	16 Growth and Form	493–528
Show Two • It Grows and Grows	94–98	17 Geometric Growth	529–568
		18 Inaccessible Distances	569–594
Show Three • Stand Up Conic	99–104	19 Reflecting the Universe	595–628
		20 New Geometries	629–660
Show Four • It Started in Greece	105–110	21 Symmetry and Patterns	661–692
		22 Tilings	693–722
Overview Show • Computer Science	111–112		
Show One • Rules of the Game	113–116		
Show Two • Counting by Two's	117–120		
Show Three • Creating a Code	121–123		
Show Four • Moving Picture Show	124–126		

Note: Part III of the textbook, Coding Information, pages 273–328, does not directly correspond with programs from the television series. However, Computer Science Shows Two and Three could be used as background material for Part III.

Part I Management Science
Joseph Malkevitch, York College, CUNY
Rochelle Meyer, Nasau Community College
Walter Meyer, Adelphi University

INTRODUCTION

The Management Science television programs and the accompanying textbook are intended to introduce elementary mathematical ideas that make it possible for businesses and governments to perform their functions better. You will learn not only that mathematics is at work in numerous situations of which you were unaware, but also that you can put some of these same ideas to work in situations you face yourselves. Furthermore, many of the ideas presented are based on easy to understand diagrams and do not require complicated algebra. Obviously, supporting details cannot be provided by the video. This is the job of the text. However, the shows make it possible for you to get an overview of the subject matter and if a still picture is worth 1,000 words, who is to determine the value of moving and talking images and computer graphics?

We firmly believe that mathematics cannot be a spectator sport. You no doubt learned to ride a bicycle by getting on one, and you probably had an occasional spill. The analogue here will be taking the risk of trying the exercises and having the confidence you can learn to "ride." Also, it is important to understand that mathematics is a study of patterns. You may have had the experience of being introduced to a fellow student by a friend, and found subsequently you were seeing that person all over campus. Once you see the power that abstraction and looking for patterns brings, you too will see the themes presented here at work in many unexpected places.

OVERVIEW SHOW

You are probably so accustomed to seeing bread on the shelves of the supermarket, to watching skyscrapers and houses rise, to hearing fire engines and ambulances rushing to a fire, and to seeing the garbage disappear from the cans beside your house that you have probably given little thought to how these services are provided. The delivery of goods and services involves expensive equipment and personnel. The goal in such systems is to keep the costs down and the level of services high. Companies and governments have always hired people to work on problems of this sort. Businesses have always had to solve problems with an analytic component. Inventory analysis is something that every company that makes a product has to worry about. Years ago, this used to be part of the art of being a businessman. You could save yourself a lot of money if you were skilled. Similarly, routes for trucks that make deliveries to stores were figured out by hand. Good routes could save a company a lot of money. These ad hoc methods, however, are now increasingly being replaced by formal methods and principles from the discipline of operations research. Mathematical ideas have replaced ad hoc methods so that now classes of problems can be attacked simultaneously, without attention to the details of a specific problem. For example, routing problems in general can be studied and algorithms to solve them generated. Problem solving is done rather than looking at a garbage collection problem or a snow removal problem for a specific section of a specific town. Management science sheds light on how to schedule "processors," whether these processors are doctors in an operating room, checkout clerks in a supermarket, runways at an airport, or photocopy machines in an office.

This show is designed to show the tremendous scope of applicability of the concepts of management science. Among the many ideas touched on are scheduling and organization of resources, queues (waiting lines), shortest paths and related routing problems, and various optimization problems.

SHOW ONE: STREET SMARTS: STREET NETWORKS

Many routing problems involving traversing streets in a city can be solved using a geometric tool called a *graph*. Given the portion of a map shown in **Figure 1 (a)** below, one can use dots called *vertices* to represent the corners, and line segments called *edges* to represent the streets joining two corners. This gives the graph in part (b) of **Figure 1.**

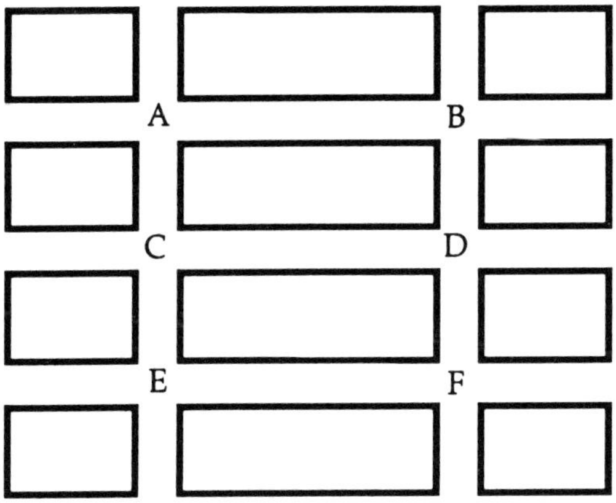

Figure 1 (a).

Figure 1 (b).

If, instead of streets, sidewalks are to be represented (modeled), the resulting graph would be:

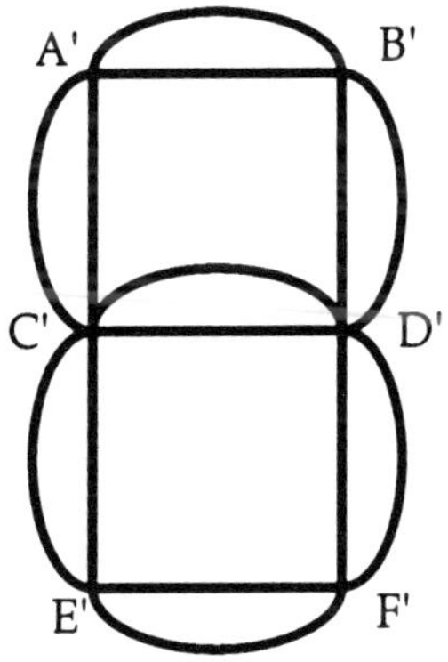

Figure 1 (c).

For the purpose of delivering mail or collecting coins from parking meters, it is desirable to start at a corner, traverse each stretch of sidewalk once and only once, and return to the start vertex. The problem of finding such a route in a graph consisting of one piece is called finding an Euler circuit. It is not hard to shown that if a graph is connected (one piece), it has an Euler circuit if all its vertices are even-valent, that is, have an even number of edges which meet at the vertex. (The graph in **Figure 1(c)** has this property.) In practice, it will rarely happen that the graph representing a street network is connected and even-valent. Hence, one would like to traverse the edges of the network starting and ending at the same vertex, with a minimum number of repeated edges. This new problem is known as the "Chinese Postman Problem," and is associated with Meigu Guan (Peoples Republic of China) who first studied the problem in detail. This problem can be solved by duplicating a minimum number of edges in the original graph so that the new graph has only even-valent vertices, a process called "Eulerizing" the original graph. It is not hard to see that the number of repeated edges is at least the number of odd-valent vertices divided by 2. There is no guarantee, however, that these few edges can be repeated. In some graphs (**Figure 2** below), every edge must be repeated to Eulerize the graph.

Figure 2.

The power of the ideas described here lies in the large number of settings where ideas of traversing the edges of a graph can arise. These include delivering mail, inspecting curbs, sweeping streets, sanding streets after a snowstorm, and many others.

Skill Objectives

1. Learn the graph concept.
2. Learn to use graph models for simple problems in operations research.
3. Learn to find valances of vertices of a graph.
4. Be able to find Euler circuits in a graph.
5. Be able to solve Chinese postman problems by Eulerizing a graph.
6. Be able to find situations modeled by edge traversal in a graph.

Self-test

MULTIPLE CHOICE

1. In the graph below the valance of vertex A is
   a. 2
   b. 4
   c. 6
   d. none of the above

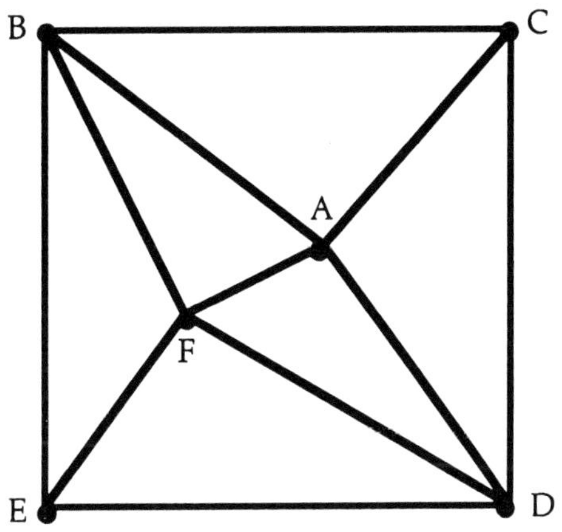

2. The graphs below which are *not* connected are
   a. G_1, G_3
   b. G_1, G_2, G_4
   c. G_3
   d. G_2, G_3

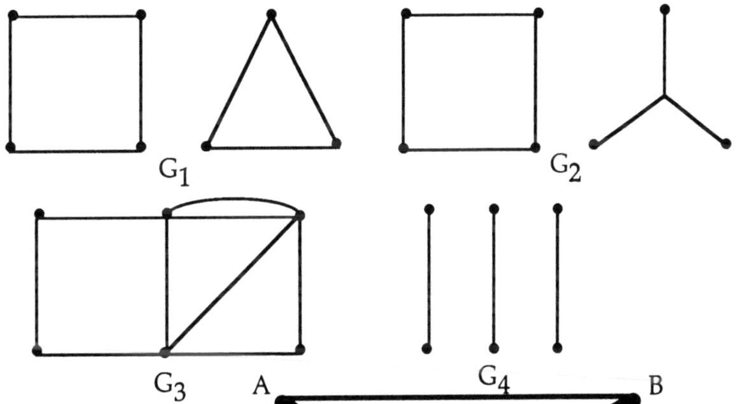

3. The graph below
   a. has an Euler circuit.
   b. is not connected.
   c. has an odd-valent vertex.
   d. has only vertices of valence 6.

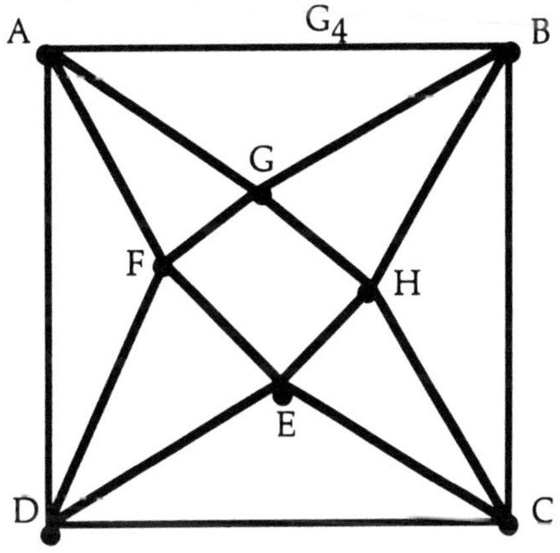

4. An Euler circuit for the graph below is
   a. A B C J H I C D E F H G A
   b. A B C I H G A
   c. J C D E F H A
   d. J C B A G H I C J

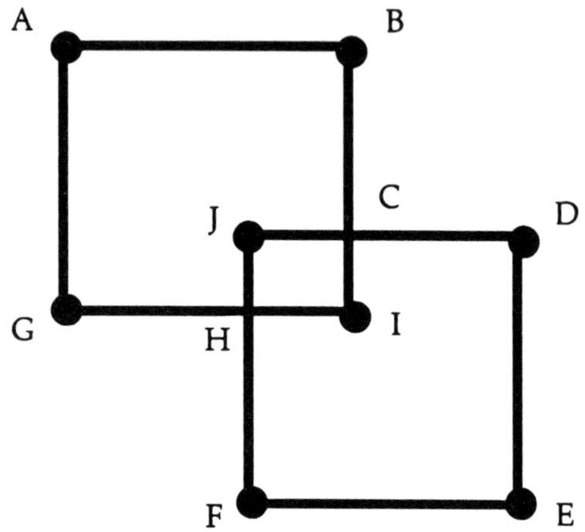

5. Determining if a graph has an Euler circuit might arise from a problem which models
   a. assigning workers to the jobs for which they are best qualified.
   b. scheduling operations.
   c. snow removal for a portion of a city's streets.
   d. inspecting traffic lights at the corners of a small village.

6. Since the graph below has 6 odd-valent vertices, to Eulerize the graph requires the duplication of at least
   a. 1 edge.
   b. 2 edges.
   c. 0 edges.
   d. 3 edges.

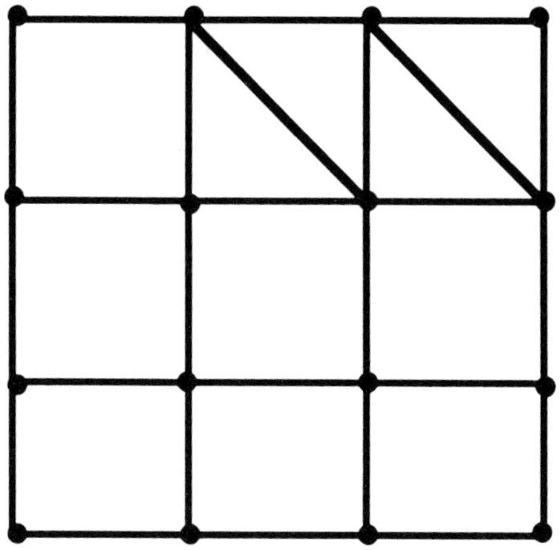

7. The minimum number of duplications to Eulerize the graph in **Problem 6** is
   a. 1 duplication.
   b. 3 duplications.
   c. 4 duplications.
   d. 7 duplications.

8. A graph which has vertices of valence 4, 3, 2, 2, 1 is

a.

b.

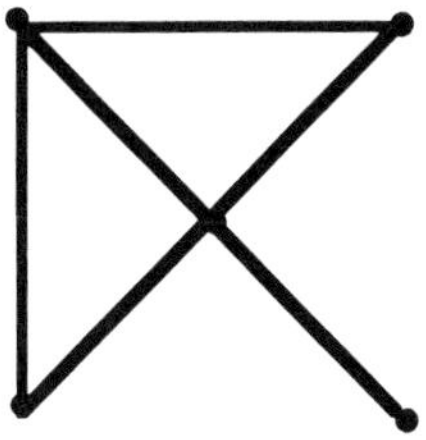

c.

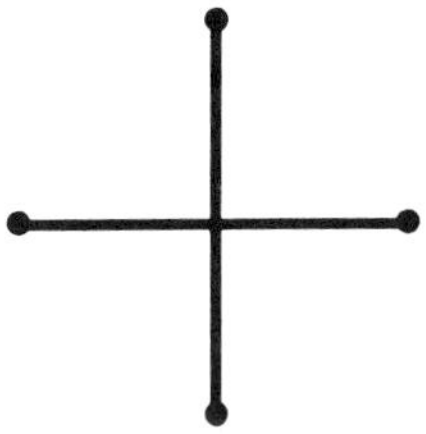

d.

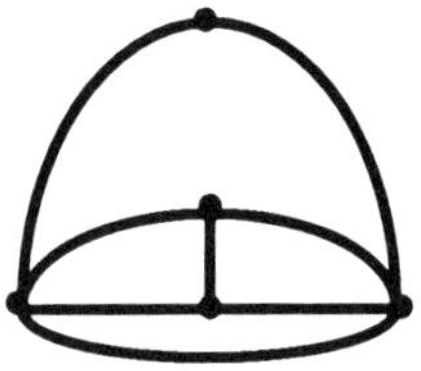

9. A route which traverses each edge at least once with a minimal number of repetitions
is
   a. A, B, D, E, D, C, A
   b. B, C, A, B, D, E, D, B
   c. A, C, E, D, C, B, A
   d. A, B, C, D, E, D, B, C, A

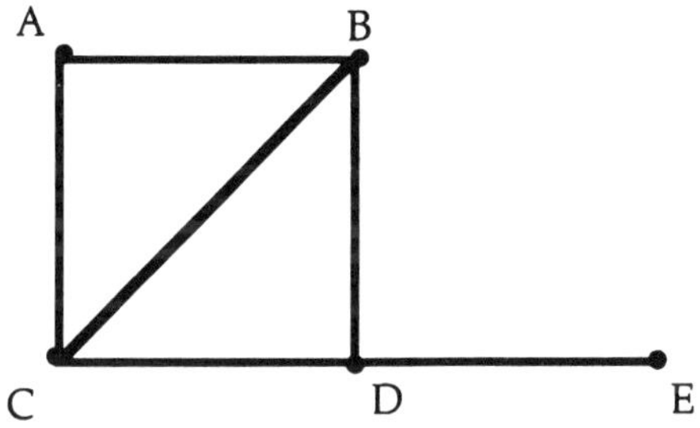

10. A graph G which is connected has an Euler circuit if
   a. it has 20 edges.
   b. it has 10 vertices.
   c. it has no circuits.
   d. it has vertices all of valance 4.

ANSWERS:
1. (b), 2. (b), 3. (a), 4. (a), 5. (c), 6. (d), 7. (c), 8. (b), 9. (d), 10. (d).

Sample Problems

11. Write down an Euler circuit for the graph below or give a reason why it has none.

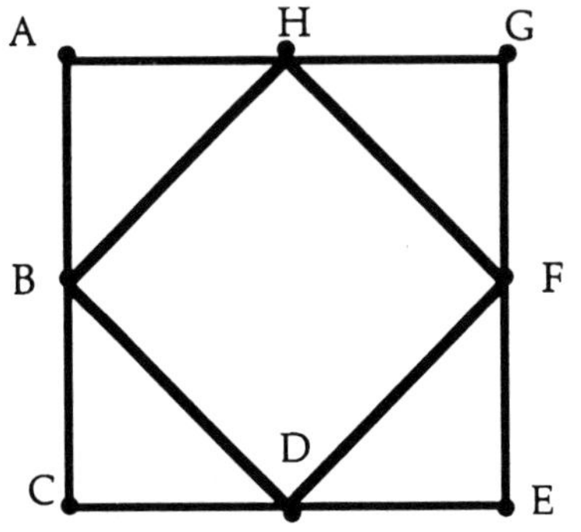

12. Draw a graph which is connected and has valences 5, 4, 3, 3, 1.

13. Draw a graph which is not connected, and has vertices of valences 2, 2, 2, 2, 3, 3.

14. Solve the postman problem for the graph shown.

ANSWERS

11. A, B, H, F, D, B, C, D, E, F, G, H, A

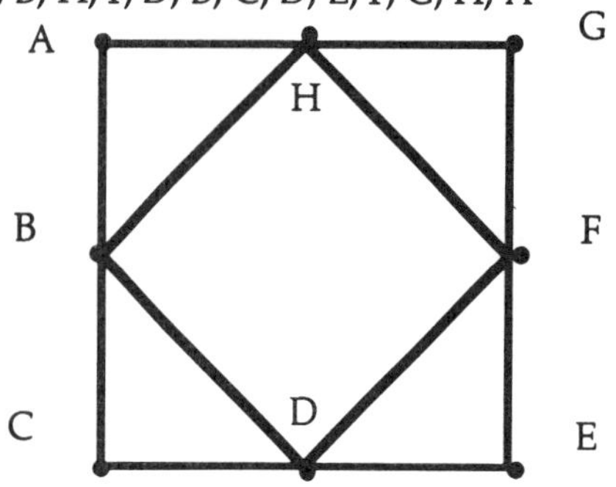

12. There are many possible answers; here is one:

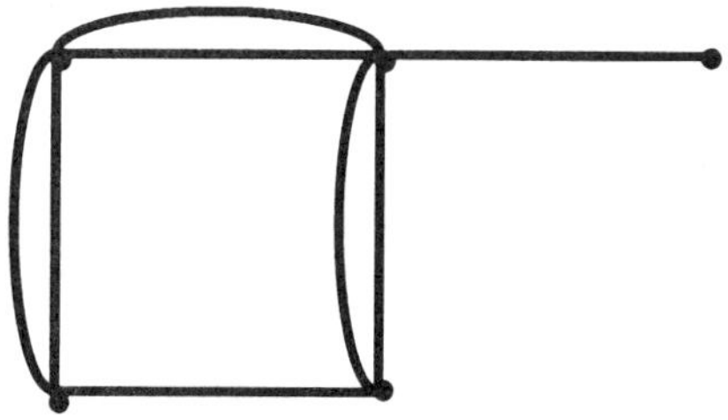

13. Here is one possible solution:

14. The graph with the duplicated edges shown below is a solution with four repeated edges:

SHOW TWO: TRAINS, PLANES, AND CRITICAL PATHS

Many problems faced by governments or businesses require the routing of either vehicles or people to achieve a specific goal. For example, a parcel post service wishes to have a truck deliver and pick up as many packages as possible from different sites in one working day. After a hurricane, a sewer inspection must be carried out at the corners in a small village to make sure that local flooding due to clogged sewers has not occurred. Both of these problems can be modeled as requiring a route which visits each *vertex* of a graph once and only once. This contrasts with the previous program where we were concerned with routes which visited each *edge* of a graph once and only once. Although mathematicians and computer scientists have been able to construct computationally efficient (i.e., fast) algorithms for most optimization problems which involve traversing all the edges of graphs, no such algorithms have been found for many vertex traversal problems. In fact, it is generally believed that no computationally efficient methods for these problems will ever be found.

The best known problem involving visiting the vertices of a graph is the Traveling Salesman Problem (TSP): Given a collection of cities, each represented by the vertex of a graph, and for each pair of cities, given the cost of traveling between these two cities, find a route visiting each city exactly once and having a minimal cost.

Finding all possible tours, and picking the cheapest tour are not computationally efficient, since the number of tours,

$$\frac{(n-1)!}{2}$$

grows so rapidly with n that even the fastest computers could not solve the problem by this method. (Recall that 10! means $10 \cdot 9 \cdot 8 \cdot 7 \cdot 6 \cdot 5 \cdot 4 \cdot 3 \cdot 2 \cdot 1$, which gives the number of tours of 11 cities as about 1.8 million.)

Since TSP's occur that must be solved, mathematicians have developed approximation algorithms which quickly give what are usually relatively good solutions. One such algorithm, which is very easy to understand, requires that one go from one's current location to that city not already visited which is as close as possible. This algorithm is called *nearest neighbor*.

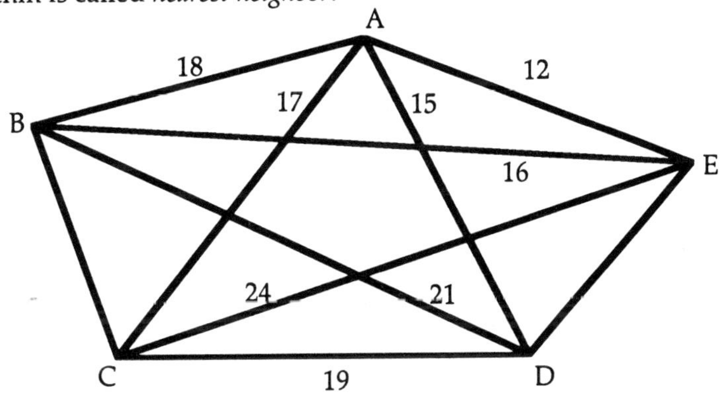

Figure 3.

For example, starting at vertex A, the nearest neighbor algorithm gives A, E, B, C, D, A, which has cost 12 + 16 + 14 + 19 + 15 = 76. It is not that easy, even for this small problem, to determine whether or not this is the cheapest tour for this problem. It is not hard to find examples, even for only four-city problems, where the nearest neighbor algorithm does not find the optimal TSP tour. Fortunately, mathematicians have found reasonably fast algorithms for the relatively large TSP's that occur in practice.

Another important operations research problem occurs when one tries to create links between vertices in a graph so that every pair of vertices is joined by a path, and the total cost of adding the links is as cheap as possible. This problem is known as the minimum cost spanning tree problem, and it frequently occurs in problems involving telephone traffic rates. This time, an easy to describe algorithm--at each stage add the cheapest edge not forming a circuit (see **Figure 4**) with edges already chosen--yields an optimal solution. This algorithm is called the *greedy algorithm* since at each stage it tries to add the

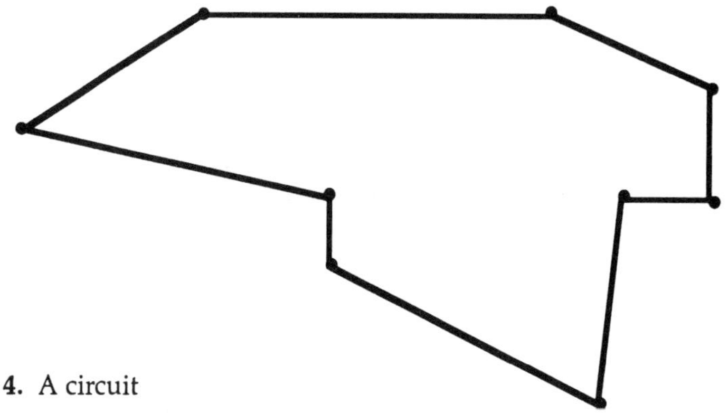

Figure 4. A circuit

cheapest possible additional edge. Although Joseph Kruskal, a mathematician at AT&T Bell Laboratories has shown this algorithm to be an optimal one, the corresponding greedy algorithm for the TSP is shown by example in the video program not to be optimal. This illustrates the need for proof of the correctness of a procedure for all situations, since otherwise, one might have been "lucky" in the few examples one tried. One of the most pervasive applications of Management Science has been helping complete complex projects as early as possible. To build a large skyscraper or to efficiently utilize airplanes requires careful planning and articulation of the tasks which make up the job. If each task is modeled as a vertex, and two vertices are joined by an arrow--directed edge--if one task precedes the other, one obtains a model for the job called an *order requirement directed graph (digraph)*. Since each task has a time associated with it, one can find the time necessary to complete a sequence of tasks by adding up the times of these tasks along a path in the order requirement digraph. Although it may not seem intuitive, the earliest completion time for all the tasks must be the length of the *longest* path in the order requirement digraph. This follows from the fact that for *every* path before a task in that path can be finished, the tasks prior to it in the path must have been done. Hence, the earliest completion time is at least as large as the length of every path, and thus is equal to the length of the longest path. The longest path in an order requirement digraph is called the *critical* path since it tells those tasks which are "bottlenecks" for completing the job earlier. Reducing the time to complete tasks not on the critical path will not speed completion of the whole job. Furthermore, even shortening the time to do tasks on the critical path may not shorten job completion time if *new* critical paths are created due to shortened task times. Critical path analysis has been an invaluable tool in speeding project completion and maintaining reasonable cost levels on large projects.

Skill Objectives

1. Find approximate solutions to the traveling salesman problem using the nearest neighbor algorithm.
2. Find approximate solutions to the traveling salesman problem using the greedy algorithm.
3. Distinguish between real-world problems which are modeled by Euler circuits (or the Chinese Postman problem) and those that are modeled by Hamiltonian circuits (or the TSP).
4. Distinguish between algorithms which yield optimal solutions and those that give nearly optimal solutions.
5. Find minimum cost spanning trees by the greedy algorithm (Kruskal's algorithm).
6. Using the critical path method, find the earliest possible completion time for a collection of tasks.
7. Understand the difference between a graph and directed graph model.

Self-test

MULTIPLE CHOICE

1. The nearest neighbor algorithm for solving the TSP (Traveling Salesman Problem)
   a. never gives an optimal solution.
   b. always gives an optimal solution.
   c. is not guaranteed to give an optimal solution but for many problems gives reasonably good solutions quickly.
   d. always gives an optimal solution if there are five or fewer cities.

2. The greedy algorithm sorted-edges for solving the TSP
   a. always gives the same solution as the nearest neighbor algorithm.
   b. never yields optimal solutions.
   c. is easy to use even for large numbers of cities.
   d. is optimal for four-city problems.

3. For a six-city TSP, the number of distinct tours is
   a. 6
   b. 6!
   c. (5!)/2
   d. (6!)/2

4. Which of the following is not a variant of the TSP problem?
   a. collecting coins from pay telephone booths.
   b. painting a line down the center of the streets of a town with only two-way streets.
   c. collecting lobsters from traps in different areas of a bay.
   d. inspecting traffic lights at intersection corners.

5. What is the cost of the minimum cost spanning tree for the graph below?

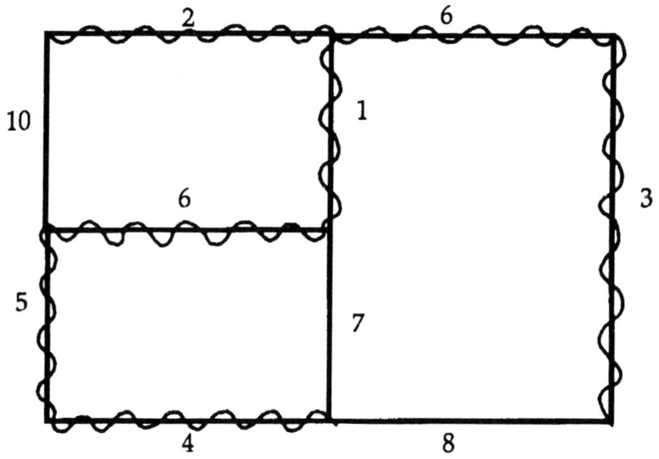

   a. 27
   b. 38
   c. 20
   d. none of the above.

6. The nearest neighbor solution to the TSP below starting at vertex A is

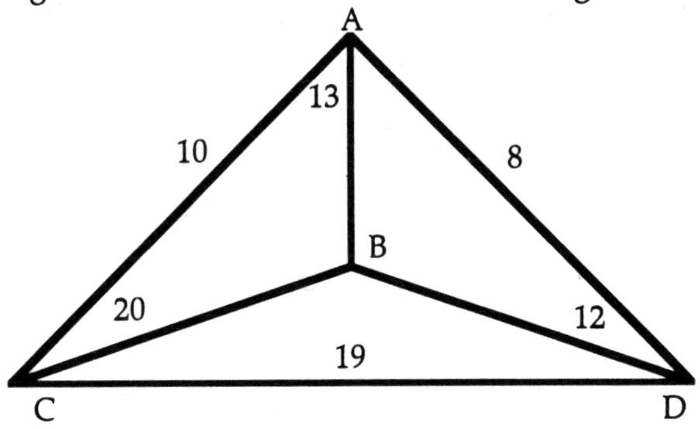

a. A D B C A
b. A D C B A
c. A D B C B A
d. A C D B A

7. The greedy algorithm for the TSP in **Problem 6** yields the following tour:
a. A B D C A
b. A B C D A
c. A D B C A
d. none of the above.

8. The critical path in an order requirement digraph for scheduling a collection of tasks
a. is the shortest length path in the digraph.
b. is always unique.
c. is a longest length path.
d. can never have more than four edges.

9. If a task on a critical path in an order requirement digraph has its time reduced,
a. the completion time for the job consisting of all the tasks decreases.
b. there must be a new critical path with shorter length.
c. the time necessary to complete all the tasks as early as possible may not change.
d. the earliest completion time for the tasks increases.

10. The critical path in the order requirement digraph below is:

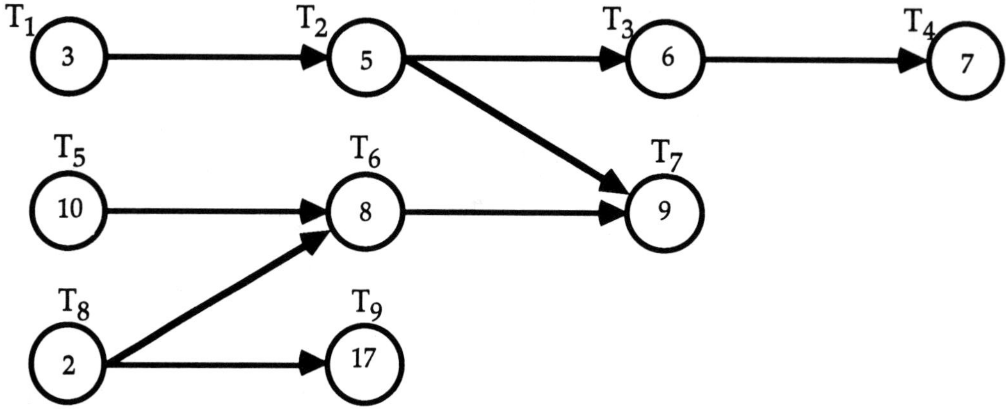

a. T_5, T_6, T_7
b. T_1, T_2, T_3, T_4
c. T_8, T_9
d. T_1, T_2, T_7

ANSWERS
1. (c), 2. (c), 3. (c), 4. (b), 5. (a), 6. (a), 7. (c), 8. (c), 9. (c), 10. (a).

Sample Problems

11. (a) Use a tree diagram to enumerate all the TSP tours starting at vertex A for the graph below.

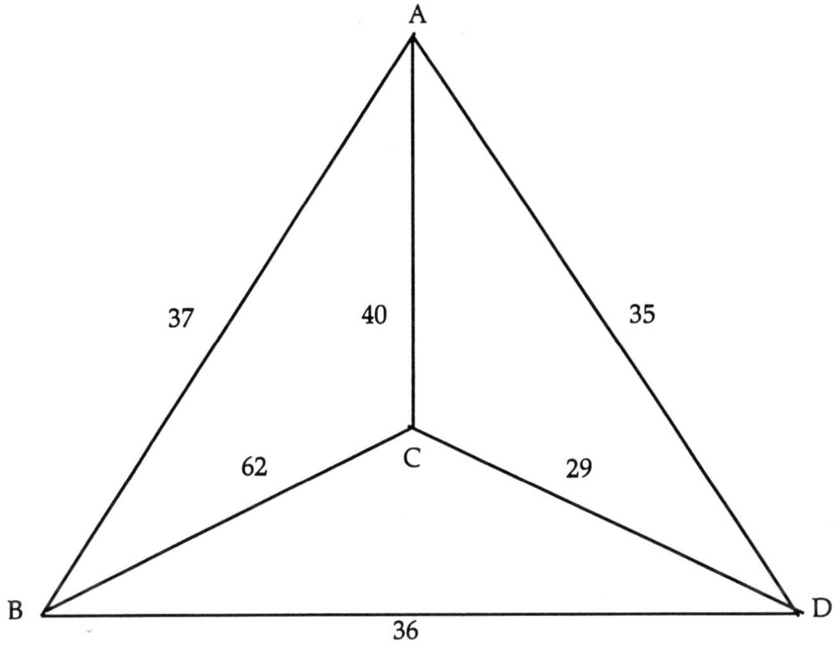

(b) Find the solution (i.e., minimum cost tour) for the TSP above.

12. Verify that the total number of distinct TSP tours for **Problem 1** is correct using the counting formula derived in the video show.

13. If the salesman in **Problem 1** wishes to visit B immediately before of after A, how many different tours are possible?

14. Solve **Problem 1** by the nearest neighbor method, starting at A, B, C, and D in turn. Do you get the same solution in each case?

15. Solve the TSP problem in **Problem 1** by the greedy method. Is the solution optimal?

16. Find a minimum cost spanning tree for the graph below:

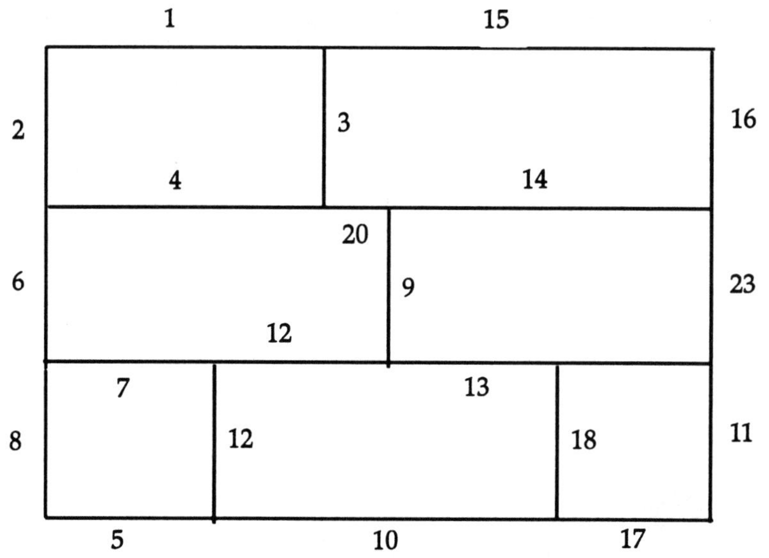

17. Find the earliest time necessary to complete all the tasks in the order requirement digraph below:

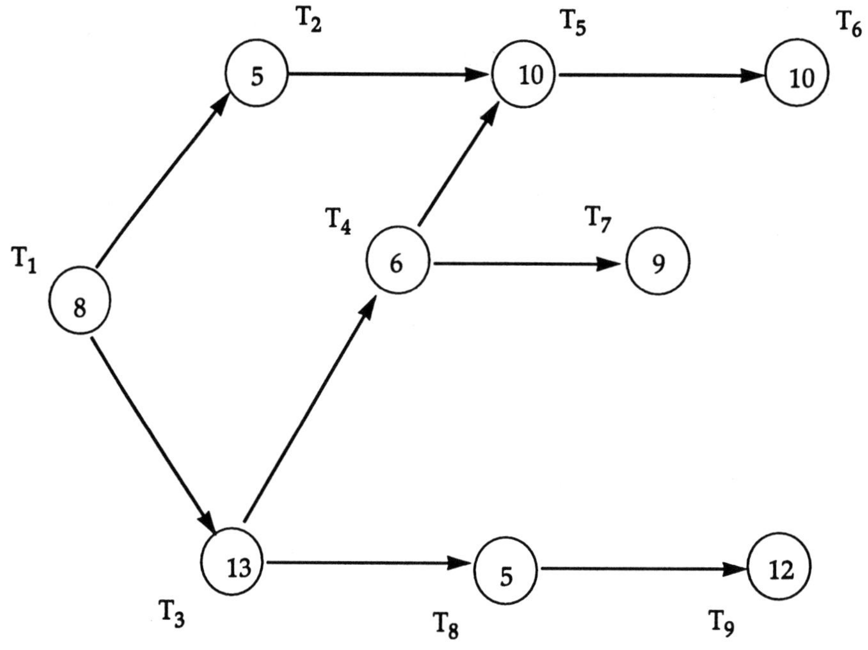

18. For the order requirement digraph problem in 7, find the critical path. If task T_2 is shortened to take 3 time units, find the earliest completion time. If task T_4 is shortened to 3 time units, find the earliest completion time.

ANSWERS:

11. (a) Only 3 of these 6 tours are actually distinct.

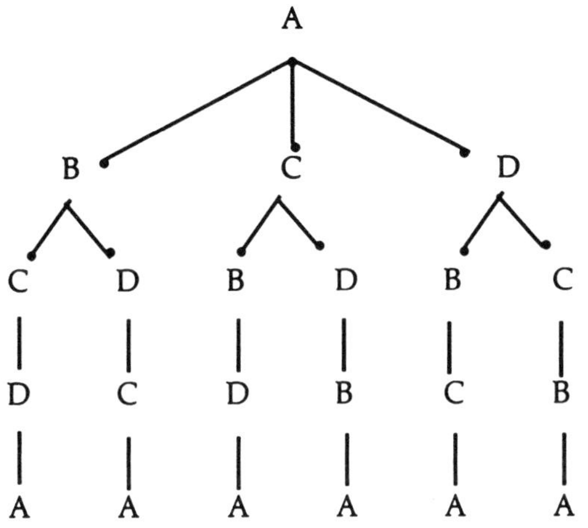

(b) A, B, D, C, A has length 142.

12. $\frac{(4-1)!}{2} = \frac{3!}{2} = \frac{6}{2} = 3$

13. The tours are:

A, B, C, D, A
<u>A, B, D, C, A</u>

A, D, C, B, A
(This last tour is the same as the first.)

14. A: A, D, C, B, A
B: B, D, C, A, B
C: C, D, A, B, C
D: D, C, A, B, D

15. C, D, A, B, C (not optimal)

16. This tree is generated by the Kruskal's algorithm. The cost of the tree is the sum of the weights on it edges.

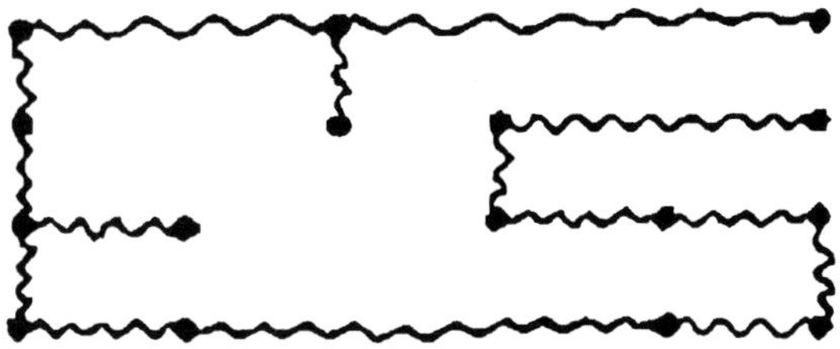

17. 47

18. If task T_2 is shortened, the completion time is unchanged. If task T_4 is shortened to 3 time units, the critical path is unchanged and the completion time would be 44.

SHOW THREE: JUGGLING MACHINES: SCHEDULING PROBLEMS

Scheduling machines and people to provide the services we depend on has become very important. Nurses, operating rooms, subway conductors, police patrol cars, airplanes, and machines in a factory all must be scheduled. To find optimal solutions to these problems, the best possible schedules, is beyond the capability of even the fastest computers using the current state of our knowledge. Many mathematicians believe no computationally fast methods will ever be found to solve these problems.

Simplified scheduling problems have been constructed to get insight into the behavior of scheduling processors. These problems obey the following simplified assumptions:

1. All the processors are identical. Any task can be done on any processor.
2. The tasks cannot necessarily be done in any order, but there is a natural precedence ordering for the tasks that can be represented by a graph with arrows on the edges (a digraph or directed graph). T_i • ⟶ • T_j means task T_i must be done before T_j can be started. This digraph is called the *order requirement digraph.* A sample order requirement digraph for "turning around" a shuttle plane from N.Y. to Washington is shown in **Figure 5** below, with the times for each task inside the vertices.

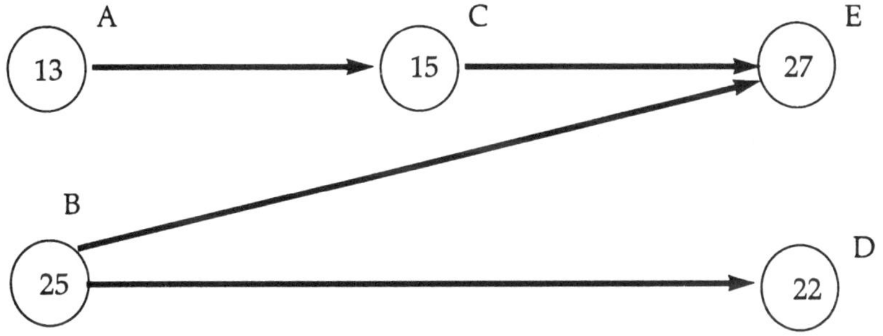

Figure 5.

3. Once a task is begun on a processor, it continues working on this task until the task is completed.

4. A processor does not start voluntarily idle. If one or more tasks is ready (i.e., all its predecessors are done), then an idle machine must begin work on a ready task.

5. There is a list showing the desired order (independent of precedence) in which the tasks might be worked on.

A task is called *ready* at a given time if all its predecessors are finished. A simple algorithm, called the *list processing algorithm,* can be used to construct schedules that incorporate these assumptions:

> At a given time, schedule the first unassigned ready task on the list, on the lowest-numbered idle processor.

We are interested in completing all the tasks at the earliest time possible. For example, the tasks in the order requirement digraph below (**Figure 6**),

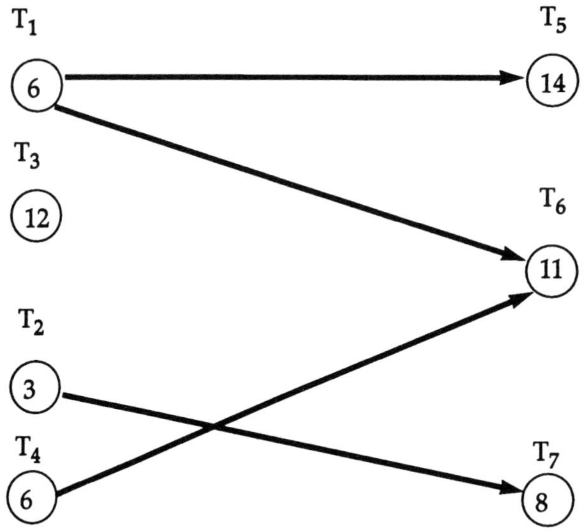

Figure 6.

when scheduled on three processors with the list (T_1, T_3, T_2, T_4, T_5, T_6, T_7), gives

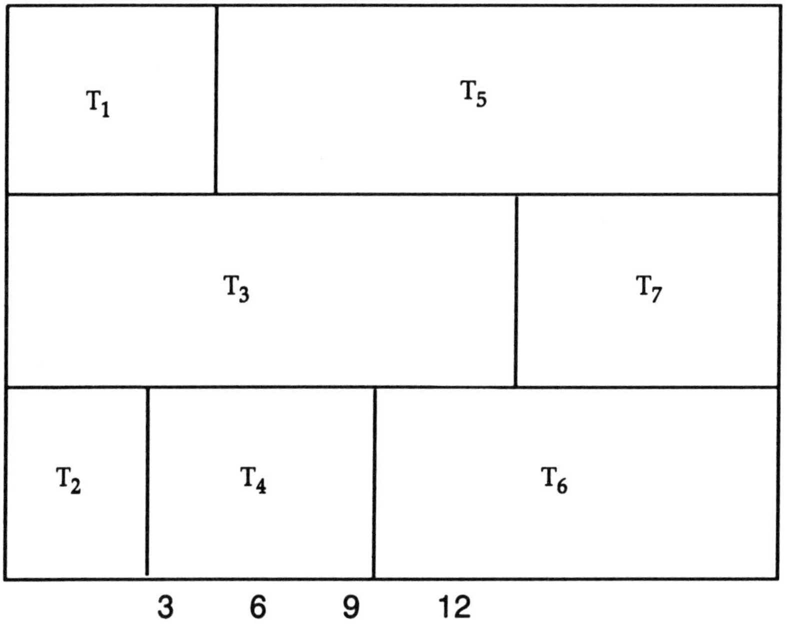

while the list (T_1, T_2, T_3, T_4, T_5, T_7, T_6) on three processors, gives

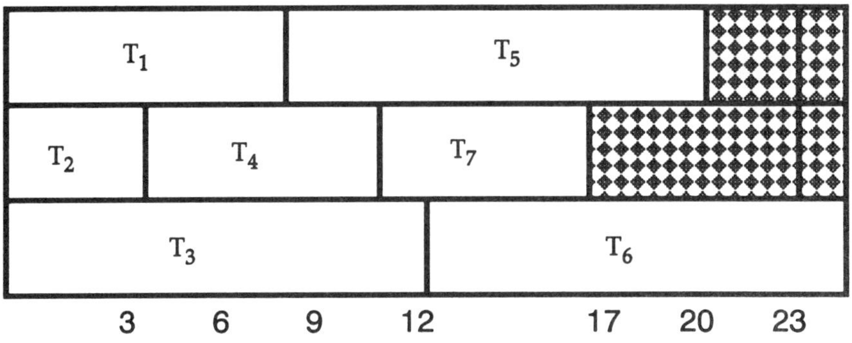

Obviously, the first schedule is optimal. Different lists will usually yield different completion times. Finding the list which gives the optimal solution can be a bit like finding a needle in a haystack.

The video show highlights some of the advantages of making a mathematical model, spelling out assumptions, and making simplifying assumptions. One advantage is that one can carefully study the range of consequences of the model. In the case of scheduling problems, the model allows unintuitive, some might say, paradoxical behavior. Thus, keeping all features of the scheduling problem fixed, except for, say, increasing the number of processors, can increase the completion time for the job and consists of all the tasks. Similarly, one can fix all factors except one of

a. loosen order requirement digraph, or
b. diminish task times,

and yet increase completion time.

Experimentation with the model adds to our understanding of both the model itself and scheduling in general. Mathematical variants, leading to further applications, grow out of the model. For example, we can examine the consequences of, assuming the order requirement digraph has no edges, that all the tasks are independent. This naturally leads to standing the initial goal of scheduling on "its head." Instead of fixing the number of processors and trying to finish all the jobs by as early a time as possible, one can fix the completion time (at a time longer than or equal to that of any tasks) and ask how many machines are needed to finish all the tasks by the fixed time. This is known as the bin packing problem because it can be thought of as finding the minimum number of bins into which one can pack a collection of given weights. Bin packing problems arise in scheduling, placing ads in fixed length time slots, and cutting insulation from fixed-length rolls of fiberglass.

There are six easy heuristic (fast to implement but not necessarily producing an optimal answer) algorithms for bin packing: Given a list of weights $w_1, \ldots, w_n$ to pack into bins of size W

a. Next Fit (NF)

Put the next weight into the current bin.
If it won't fit start a new bin; close the current bin permanently.

b. First Fit (FF)
Put the next weight into the first open bin into which it will fit.
If it fits in no open bin, start a new bin.

c. Best Fit (BF)
Put the next weight into that open bin with the most room left. If it fits in no open bin, start a new bin.

There are also versions of these three algorithms where the weights are first listed in decreasing order. These algorithms are known as next fit decreasing (NFD), first fit decreasing (FFD), and best fit decreasing (BFD).

It is known, for example, that first fit decreasing always produces a number of bins no worse that 11/9 (optimal number of bins) + 3.

Here is a simple example of how these six algorithms NF, FF, BF, NFD, FFD, and BFD work on the weights 1, 1, 9, 7, 1, 4, 2, 3, 5, 3, 3, 4 in bins of size 10.

NF

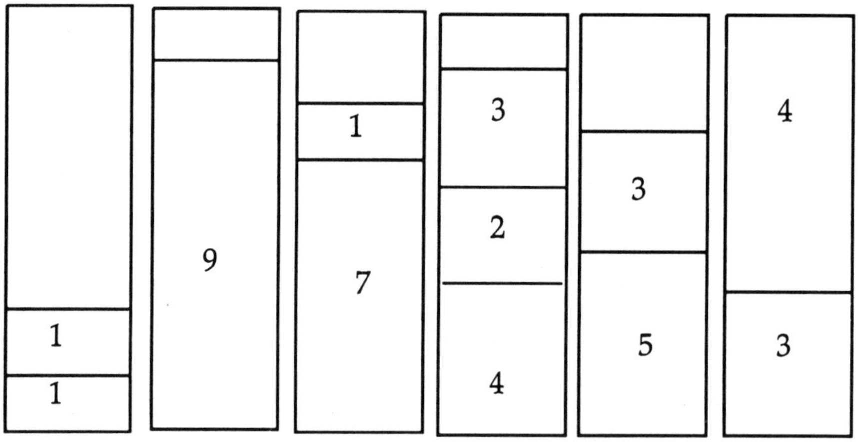

FF

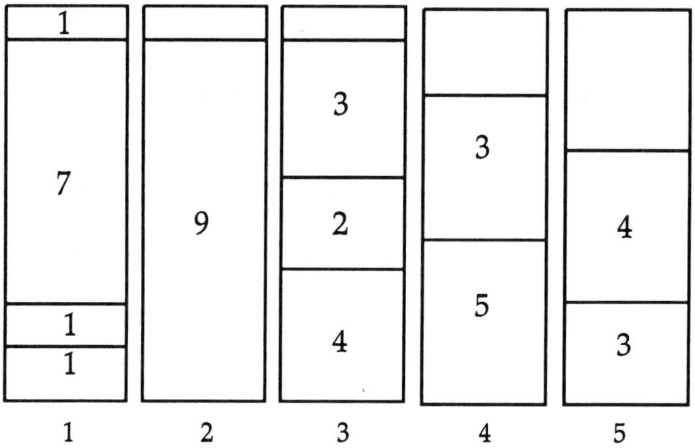

BF

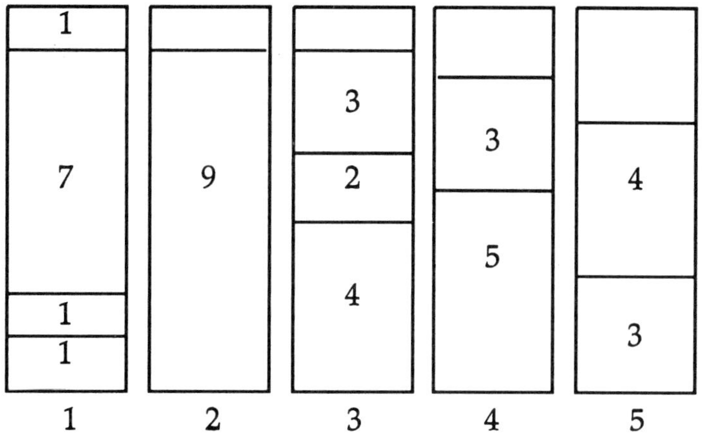

For the decreasing versions, one list used is: 9, 7, 5, 4, 4, 3, 3, 3, 2, 1, 1, 1

NFD

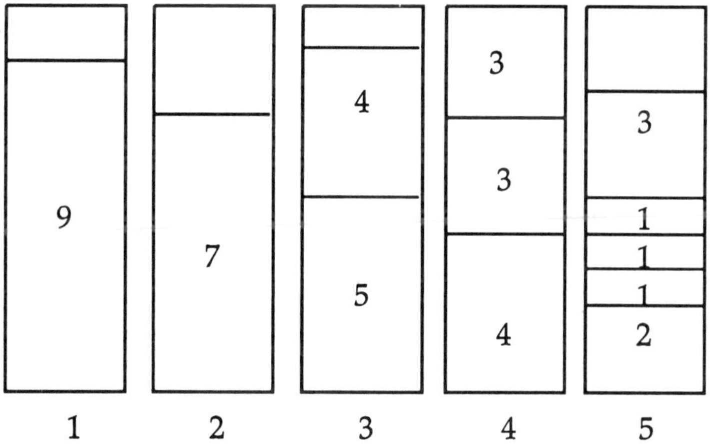

FFD

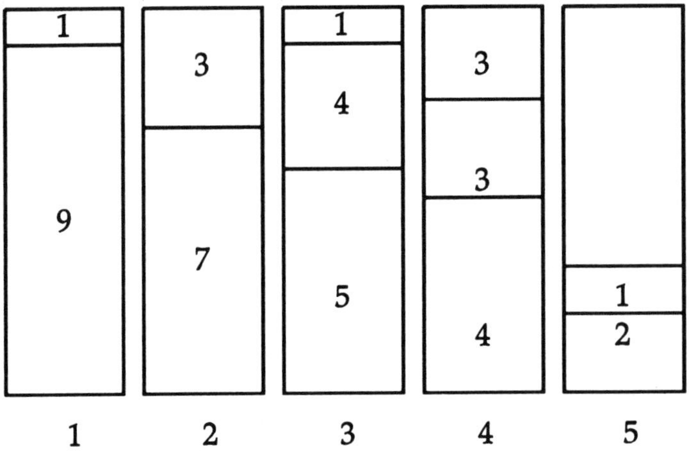

BFD

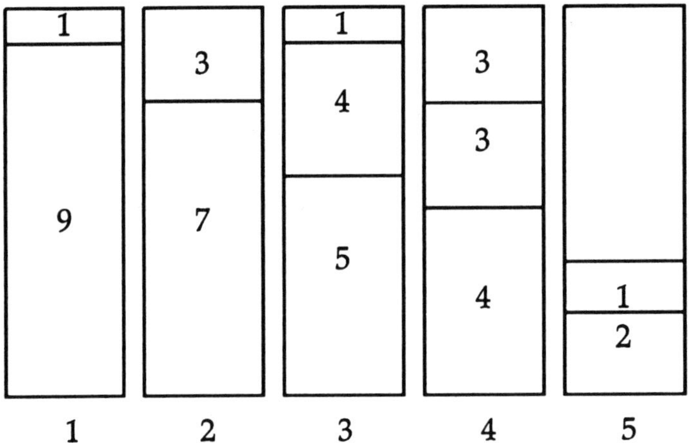

(Note that although the last two packings look identical, the weights were not put in position in the same order.)

Skill Objectives

1. Use the list processing algorithm to schedule tasks on identical processors.
2. Recognize situations appropriate for modeling as scheduling problems.
3. Put the critical path method together with the list processing algorithm to help in determining if a schedule might be optimal.
4. Recognize situations appropriate for modeling by bin packing.
5. Solve bin packing problems by one of six heuristic algorithms.
6. Solve independent task scheduling problems by the decreasing time list processing heuristic algorithm.
7. Understand the relationship of scheduling problems and bin packing problems.

Self-test

MULTIPLE CHOICE

1. The list processing schedule algorithm
   a. always produces optimal schedule for two processors.
   b. always produces optimal schedules when the tasks have equal lengths.
   c. always produces optimal schedules when the tasks are independent.
   d. none of the above.

2. Which statements do not apply to the list processing algorithm?
   a. no processor can stay voluntarily idle (i.e., it is free not to start work on a task all of whose predecessors are completed).
   b. a processor working on task T_i can interrupt its work to work on task T_j of greater importance.
   c. all processors are considered identical, that is, any of them can do any of the tasks.
   d. all tasks except one must have at least one predecessor.

3. Which statements are false for an order requirement digraph for scheduling processors?
   a. must contain at least one directed edge.
   b. cannot contain a route starting at a vertex w and returning to w following edges of the graph.
   c. cannot contain more edges than vertices.
   d. cannot be used to represent independent tasks.

4. A scheduling algorithm which schedules 20 equal-length tasks of nine units on four processors can generate a schedule where the tasks are all completed by time
   a. 20
   b. 9
   c. 45
   d. 40

5. If the order requirement digraph for a collection of tasks is:

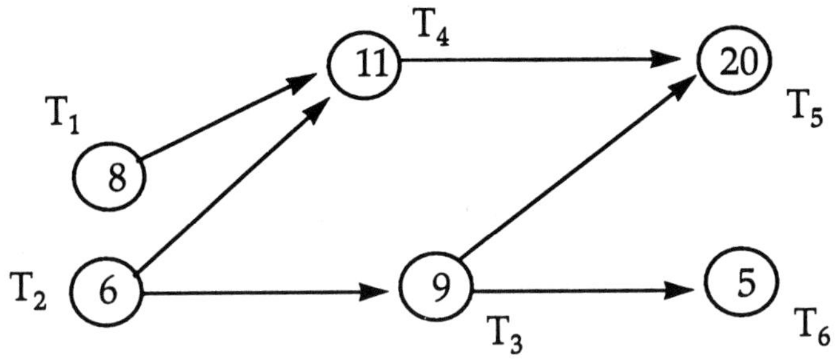

then the critical path would be
a. T_1, T_3, T_6
b. T_1, T_4, T_5
c. T_2, T_3, T_5
d. T_2, T_4, T_5

6. If, in the digraph in **Problem 5**, the time of task T_1 is reduced to 3, the earliest completion time for the tasks is
a. 37
b. 20
c. 34
d. 35

7. If, in the digraph in **Problem 5**, the time of task T_3, is reduced to 3 (other task times asthey were), then the earliest completion time for the tasks is
a. 14
b. 37
c. 29
d. 39

8. The bin packing problem requires
a. the largest weight to be packed in the first bin.
b. the smallest weight to be packed in the last bin.
c. the minimum number of bins to pack all the weights.
d. no bin to be less than half full.

9. The first fit decreasing algorithm for packing bins guarantees an answer which, in the worst case, is less than or equal to m (optimal) + 3 where m is:
a. 11/7
b. 2
c. 11/9
d. 3/2

10. If the weights 10, 6, 8, 7, 9, 4, 2, 1, 1 are to be packed using the best fit decreasing algorithm, the weights are first relisted in the order:
    a. 10, 6, 8, 7, 9, 4, 2, 1, 1
    b. 1, 1, 2, 4, 6, 7, 8, 9, 10
    c. 10, 9, 8, 7, 6, 4, 2, 1, 1
    d. 10, 1, 7, 4, 9, 2, 1, 6

ANSWERS

1. (d), 2. (b), (d), 3. (a), (d), 4. (c), 5. (b), 6. (a), 7. (d), 8. (c), 9. (c), 10. (c).

Sample Problems

11. The digraph below (times indicated within vertices) is the order requirement digraph for a collection of eight tasks.

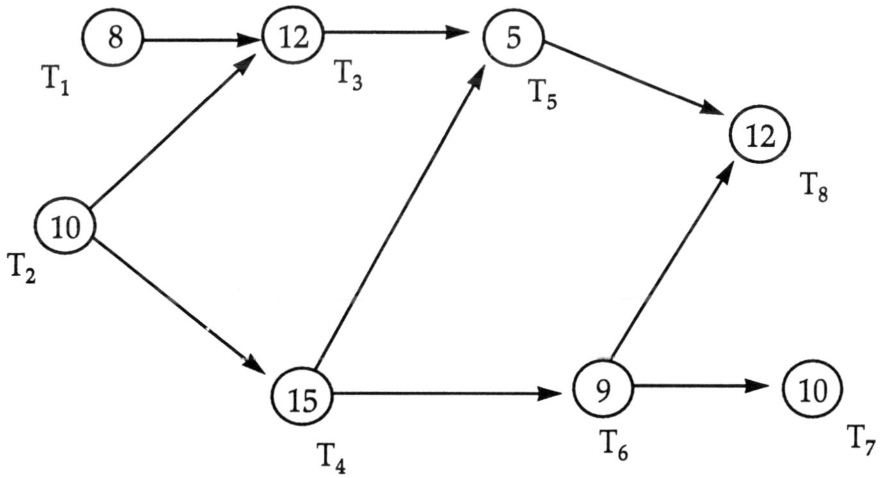

a. What is the earliest possible completion time for all the tasks?
b. List the tasks on the critical path.
c. If the time of task T_3 is reduced to 8, does this affect the earliest completion time? (If so, give new earliest completion time and associated critical path.)
d. Repeat c. if T_6 is reduced to four time units.
e. What is the completion time for these tasks on two processors using the list processing algorithm if the list used is $T_1, T_2, T_3, T_4, T_5, T_6, T_7, T_8$?
f. Is the schedule you found in e. optimal?

12. Schedule independent tasks of lengths 9, 8, 1, 6, 12, 3, 1, 9, 2, 11 on three processors, using the list algorithm and using the list corresponding to the times as ordered above.

13. Solve **Problem 2** using the decreasing time list algorithm.

14. Given the order requirement digraph below, using two different methods, give lower bounds (estimates) for the earliest completion time for all the tasks as scheduled on two processors:

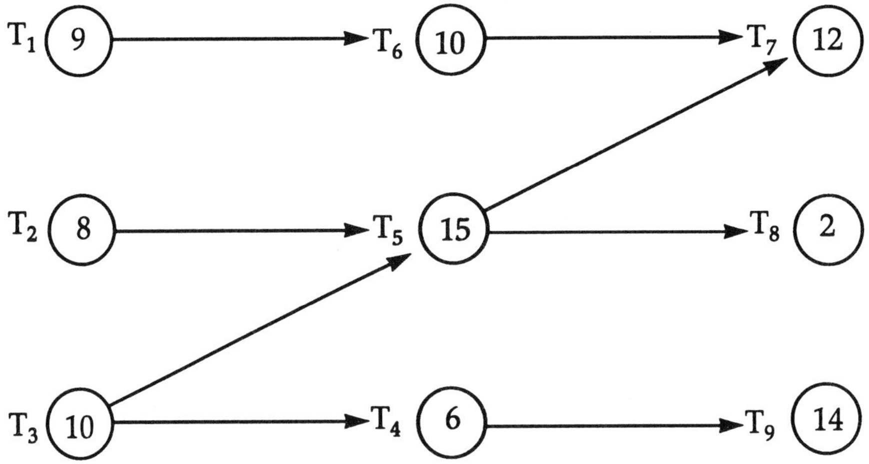

15. What would be the minimum number of typists, all having equal skill, that must be hired to type a series of projects requiring the number of minutes below within six hours?

Project times:
30, 60, 90, 60, 60, 30, 90, 90, 30, 60, 30, 30, 60, 60, 30, 30.

NOTE: We assume a typist can work six hours without interruption.

ANSWERS

11. a. 46 (length of the longest path).
    b. T_2, T_4, T_6, T_8.
    c. No. T_3 isn't on the critical path.
    d. The new critical path is T_2, T_4, T_5, T_8.
    e.

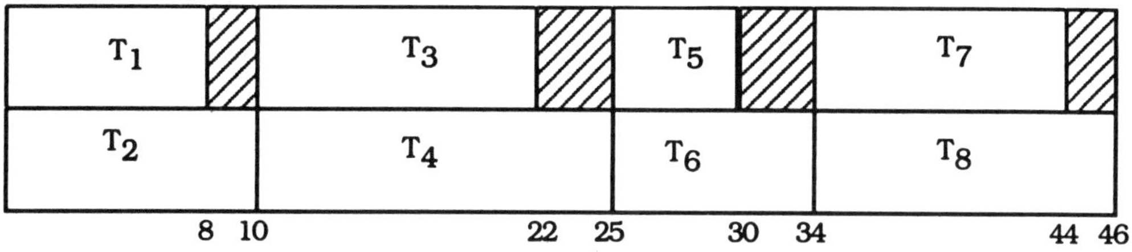

f. Yes. (The completion time equals the length of the critical path.)

12.

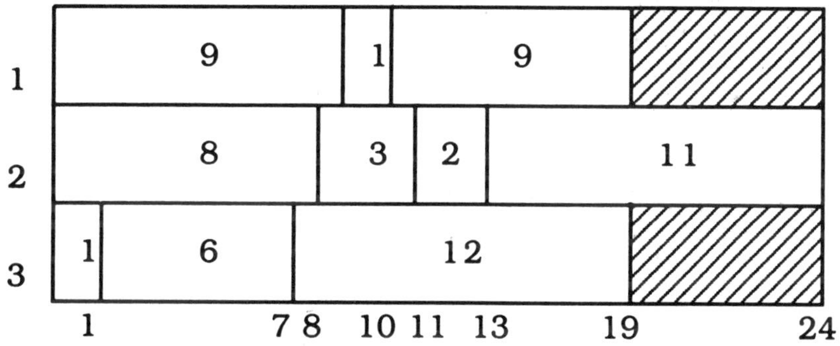

13. Listing the tasks in order of decreasing times gives the list: 12, 11, 9, 9, 8, 6, 3, 2, 1, 1. The schedule produced from this list is:

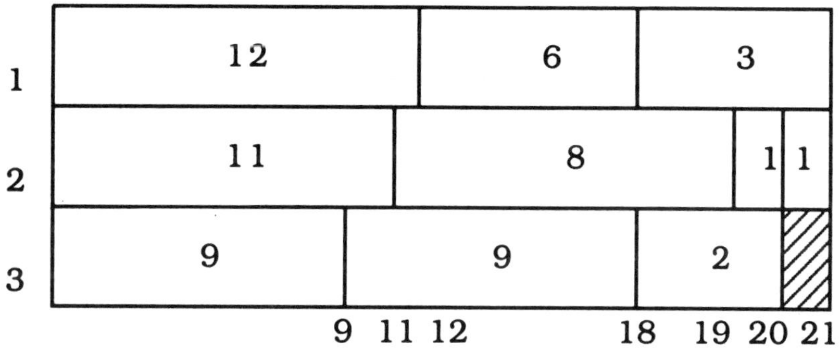

Clearly, this schedule is an optimal one.

14. a. One method to estimate the earliest completion time is to find the longest path in the digraph. This so-called critical path gives an estimate for the earliest completion time. Since T_3, T_5, T_7 is the longest path, and have length 37, completion of all tasks before time 37 is impossible.

    b. A second method to estimate the earliest completion time is to add all the task times and divide by the number of machines, two in this case. The task times add up to 86. Since 86/2 = 43, the tasks cannot be completed processors before time 43. Note, there is no guarantee that an optimal schedule for this problem will not take more than 43 time units.

15. We can interpret this problem as a bin-packing problem with bins of size 360 minutes. If we apply the first fit decreasing heuristic, we pack the weights in the order 90, 90, 90, 60, 60, 60, 60, 60, 60, 30, 30, 30, 30, 30, 30, 30, we obtain

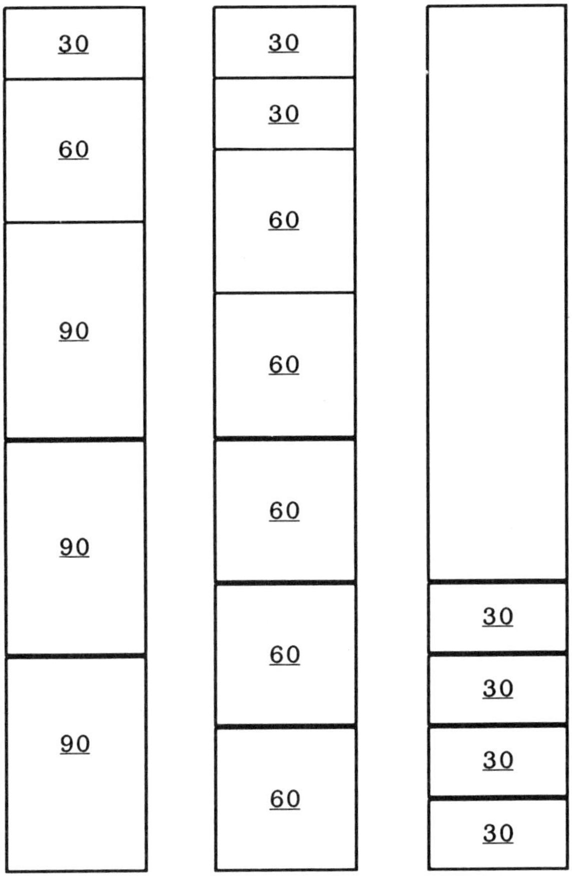

Hence, three typists should be hired. This is clearly optimal.

SHOW FOUR: JUICY PROBLEMS: LINEAR PROGRAMMING

In many situations a manufacturer has a choice of products to make. Each of these products requires specific raw materials, of which only a fixed supply is available. We can make a chart (**Figure** 7) showing this information:

	CAR	BUS	SUPPLY
Plastics	½ ton (per car)	1 ton (per bus)	45 tons
Metals	2 tons (per car)	1 ton (per bus)	60 tons

Figure 7.

The manufacturer wishes to decide what mixture of cars and buses to manufacture to *maximize* profit, where each car produces a $2,000 profit and each bus produces a $3,000 profit. Of course, one is limited to non-negative values for the numbers of cars and buses produced.

Similar problems arise in the design of optimal juice blends or the design of diets (for a hospital of day-care center) which meet nutritional requirements and minimize costs.

Problems such as these can be solved using a powerful management science tool known as linear programming. The basic ideas will be illustrated constructing a mathematical model for the cars and buses problem. Let x denote the number of cars produced and y the number of buses. Clearly x and y satisfy
$x \geq 0$, $y \geq 0$ (i.e., x and y are greater than or equal to zero), since one cannot make negative amounts of goods.

Using the table above resource constraints can be set up. For example consider the plastics available. If x cars are produced, since each car uses a ½ ton of plastic, $\frac{1}{2} \cdot x$ tons of plastic are used. If y buses are produced$1 \cdot y$ tons of plastic are used. The total plastic used for the cars and buses is $\frac{1}{2}x + y$, and this cannot exceed (i.e., is less than or equal to) the available supply 45 tons. Hence:

$$\frac{1}{2}x + y \leq 45$$

A similar inequality can be obtained for the metals:

$2x + y \leq 60$

The profit obtained, P, will be given by

$P = 1000x + 3000y$

Summarizing, we wish to maximize $P = 1000x + 3000y$ subject to
$x \geq 0$
$y \geq 0$
$\frac{1}{2}x + y \leq 45$ (*)
$2x + y \leq 60$

We can draw a visual image of the situation by recalling that one can draw pictures of equations. This process is illustrated in the figure below. The point representing 50 cars and 30 buses is labeled A (below), and A has coordinates (50, 30).

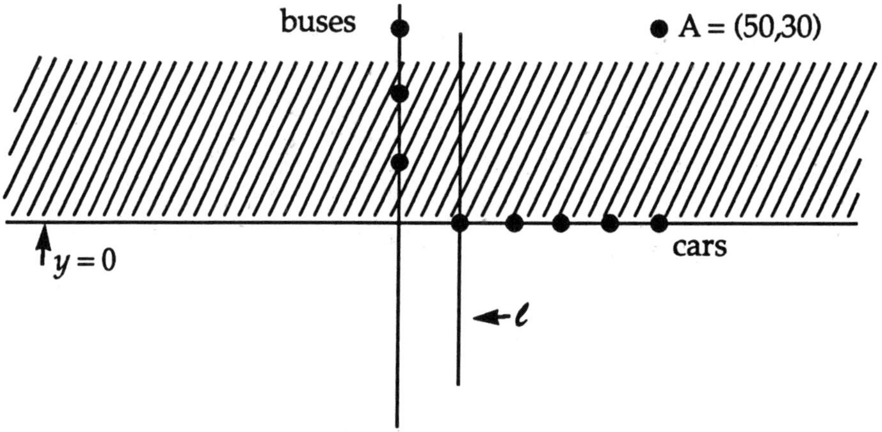

In this notation cars are always listed first. The equation $x = 10$ is represented by the vertical line ℓ, and in general equations of the kind $x + y = 40$ will represent lines and inequalities represent "half-planes." For example, $y \geq 0$ represents the shaded region above the line $y = 0$. These points are said to satisfy the inequality. In general, an inequality such as $x + 2y \leq 40$ will represent the region on one side of the line $x + 2y = 40$.

The points which satisfy all the inequalities in (*) are shown in the next figure. This region is called the feasible region for the linear programming problem. Since the feasible region has infinitely many points, how can we choose the one that maximizes the profit? It is a remarkable theorem that if the feasible region fits within some circle, then the maximum (and minimum) must occur at some corner points, and are found algebraically solving pairs of equations representing the lines which bound the feasible regions. For example,

$x = 0$ and $\frac{1}{2}x + y = 45$

are lines which meet at the corner point (0, 45), which satisfies both equations.

The complete list of corner points in this example are

A = (0, 0), B = (0, 45), C = (10, 40), and D = (30, 0).

We obtain the following values for P by substitution into $P = 1000x + 3000y$.

P at A equals 2000(0) + 3000(0) = 0

P at B equals 2000(0) + 3000(45) = 135,000

P at C equals 2000(10) + 3000(40) = 140,000

P at D equals 2000(30) + 3000(0) = 60,000

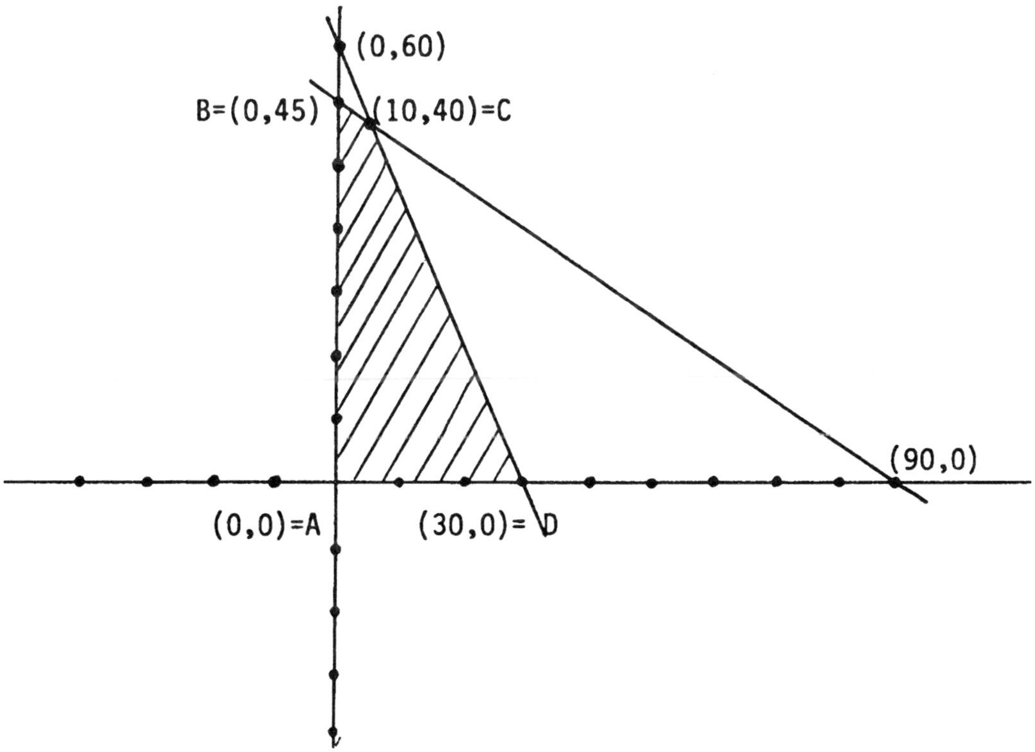

Figure 8.

Hence, the most profit is made by producing 10 cars and 40 buses. Of course, the model assumes that one can in fact sell this number of cars and buses if one adopts this production schedule.

In more realistic linear programming problems, the number of variables (product to make) may be several hundred and the number of inequalities (resource constraints) may be several thousand. Clearly, hand solutions to such problems are impossible. Using a mathematical technique called the *simplex method* developed by George Dantzig, one can solve problems of this size on a computer in a reasonable amount of time. Recently, N. Karmarkar (AT&T Bell Labs) has found a new method of great theoretical interest which may make it possible to speed up the solution of large linear programming problems.

The variety of problems attackable by linear programming techniques is staggering. Problems in scheduling, blending, and resource and manpower utilization are but a few. Mathematics has made it possible for managers to save millions of dollars and to improve the quality of goods and services, thereby improving the quality of our lives.

Skill Objectives

1. Formulate simple problems as linear programming problems.
2. Plot points, lines, and inequalities on coordinate axes.
3. Apply the corner point theorem.
4. Draw the feasible region of a linear program.
5. Understand the variety of settings in which linear programming problems arise.
6. Understand the basic concept behind the simplex method.

Self-test

MULTIPLE CHOICE

1. The table below arises from forming 1 unit of two types of fruit cocktail mixes:

	Pears	Peaches
Mix 1	3 lbs.	2 lbs.
Mix 2	4 lbs.	5 lbs.
Supply	600 lbs.	300 lbs.

If x denotes the number of units of mix 1 produced, and y denotes the units of mix 2 produced which of the following inequalities are *not* valid?
(i) $x \geq 0$, (ii) $y \geq 0$, (iii) $3x + 4y \leq 600$, (iv) $4x + 5y \leq 600$.
a. (i) and (ii)
b. (iii) and (iv)
c. (iii) only
d. (iv) only

2. Which of the following statements are true about linear programming problems?
(i) The number of variables must not exceed 8.
(ii) The number of constraints must not exceed 11.
(iii) The minimum value for a polygonal feasible region must be at a corner point.
(iv) The answer obtained to the problem can be negative.

a. (i) and (ii) only
b. (iii) and (iv) only
c. (i) and (iii) only
d. (ii) and (iv) only

3. If an industrial resource allocation problem has the property that to make 10 of an item requires A units of product z, but to make 100 of the items requires $5A$ units of product z, then
a. Karmarkar's method will help solve the problem.
b. the simplex method will run slowly on such a problem.
c. the corner point method is suited for such a problem.
d. linear programming techniques will not apply.

4. The point B in the graph below has coordinates

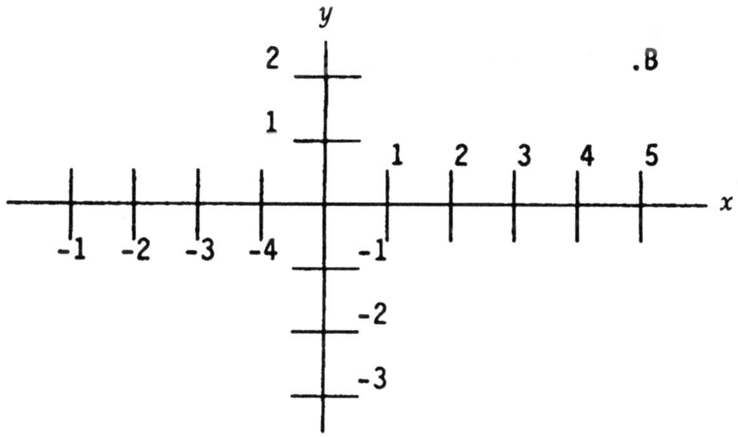

a. (2, 2)
b. (2, 5)
c. (5, 2)
d. (-1, 2)

5. The profit function, $P = 7x + 8y$, is to be minimized over a triangular shaped feasible region whose corner points are

$$A = (2, 9) \quad B = (5, 11) \quad C = (13, 1)$$

The minimum value of P is:
a. 0
b. 56
c. 15
d. 86

6. The point which satisfies both of the equation $2x + y = 8$ and $3x - y = 7$ is
a. (0, 0)
b. (2, 3)
c. (3, 2)
d. (7, 8)

7. The graph of the collection of points which satisfy $3x - 5y = 2$ is a
a. ray.
b. straight line.
c. line segment.
d. vertical line.

8. The graph of the collection of points which satisfy $x + y \geq 4$ is
a. a pair of parallel lines.
b. a pair of perpendicular lines.
c. a half-plane.
d. a straight line.

9. Linear programming
(i) can be used to solve problems involving the blending of aviation fuels.
(ii) can be used to solve problems involving the scheduling of prison guards.
(iii) cannot solve problems where optimal values occur at non-integer coordinate corner points.
(iv) is rarely used in recent years.

Of the statements above, which are false:
a. (iv) only
b. (ii) and (iv)

c. (iii) and (iv)
d. (ii) and (iii)

10. The simplex algorithm works as follows:
   (i) Starting at any corner point P, the algorithm selects next neighbor of P which gives the best improvement in the function to be optimized.
   (ii) All the corner points of the feasible region are always tested in a systematic manner to find the optimal one.
   (iii) The optimal vertex is obtained by finding any Hamiltonian circuit, and tracing it starting at any vertex.

   Of these, which are true:
   a. (i)
   b. (ii)
   c. (iii)
   d. none of (i), (ii), or (iii)

ANSWERS
1. (d), 2. (b), 3. (d), 4. (c), 5. (d), 6. (c), 7. (b), 8. (c), 9. (a), 10. (a).

Sample Problems

11. The corner points of a polygonal feasible region of a linear programming problem to minimize the cost function $C = 10x + 50y$ are:

$A = (2, 7)$ $B = (9, 6)$ $C = (3, 3)$

Find the minimum cost.

12. What are the coordinates of the points A, B, and C in the graph below:

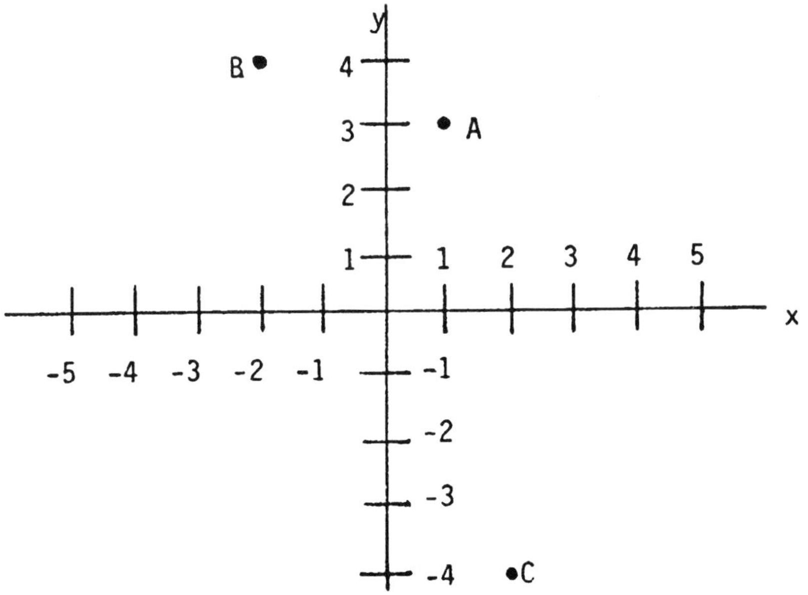

13. To build zingers and zonkers requires the use of cobalt and chrome steel as shown in the chart below. If zingers and zonkers yield profits \$100,000 and \$400,000 respectfully, set up the linear programming problem which represents this situation.

	Tons chrome steel per item	Tons cobalt steel per item
Zinger	6	4
Zonker	1	2
Supply	120	200

14. A feasible region of a linear programming problem is bounded by the lines $x + 4y = 44$ and $2x + y = 60$. Find the corner point these lines determine.

15. Draw the region bounded by the inequalities

$x \geq 0$

$y \geq 0$

$x \leq 7$

$y \leq 6$

$x + y \leq 10$

ANSWERS

11. 180
12. A = (1, 3); B = (-2, 4); C = (2, -4)
13. If x denotes the number of zingers made, and y denotes the number of zonkers, then maximized profit = 100,000x + 400,000y subject to:

$x \geq 0$

$y \geq 0$

$6x + y \leq 120$

$4x + 2y \leq 200$

14. (28, 4)
15.

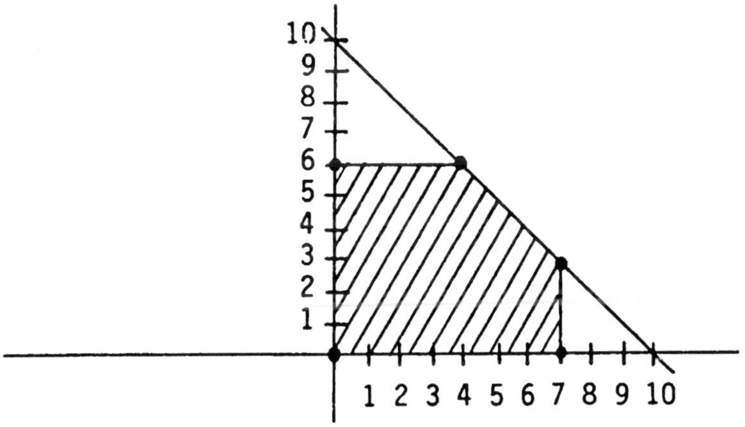

Part II Statistics
David S. Moore
Purdue University

INTRODUCTION

Numerical facts, or data, make up an increasing part of the information we need in order to understand our world. Opinion polls, market research, and government efforts to collect economic and social information all produce data. *Statistics* is the science of data--of gathering data, of putting them into clear and usable form, and of interpreting them to draw conclusions about the world around us.

The information researchers gather may be as vital as the fact that 7.2% of the American labor force is out of work, or as superficial as the fact that the average U.S. household watches television for exactly seven hours and two minutes each day. But in each case we must know how to collect accurate numbers, how to describe the resulting data briefly and clearly, and how to draw conclusions. In fact, collecting data, finding ways to display them, and making intelligent predictions based on them are the building blocks of the science called statistics.

OVERVIEW SHOW

In this show we first examine how and why data must be accurately collected by visiting the Bureau of Labor Statistics. There we see how our unemployment figures are gathered every month. We also look at how experimental data are collected by analyzing a study of doctors, whose results hold promise for reducing heart attacks and cancer mortality.

To see how statistics affects us every day, we travel to Oklahoma City and examine AT&T's quality-control efforts. And to see how statistics can have a direct effect on making money--lots of money--we spend some time in Atlantic City. There we learn how a casino can turn a game of chance into a sure profit.

SHOW ONE: BEHIND THE HEADLINES: COLLECTING DATA

Data are gathered for a specific purpose by *sampling* or by *experiment*.

In sampling, we decide what *population* we are interested in, then select a part of that population as a *sample*. We collect information about the sample, and from this data, draw conclusions about the population. Some important uses of sampling are:

Polls of public opinion,
Market research, such as TV program ratings,
Controlling the quality of manufactured products,
Government economic and social statistics.

The *Current Population Survey* is particularly important as the source of official statistics on employment, unemployment, and many other characteristics of American households.

If we are to be confident that sample data do represent the entire population, we must avoid bias in choosing the sample from the population. Personal choice by the sampler and voluntary response by persons wanting to be sampled are common sources of bias. Statistical methods use *random sampling* to avoid bias. A *simple random sample* is the simplest kind of random sample. National samples such as the Current Population Survey use *multistage samples* in which the random selection is carried out in several stages rather than all at once.

The results of random sampling would vary if the sampling process were repeated. But this *sampling variability* follows known patterns described by the laws of *probability*. The nature of the known pattern of sampling variability will be described in detail later in Section 3. Deliberate randomization in collecting data allows statisticians to use probability to say how often a sample result will fall within a given margin of error of the truth about the entire population. For example, a typical opinion poll result based on interviews with 1,500 persons has probability .95 of falling within ±3 percentages points of the result that would be obtained by interviewing the entire population. Larger samples give smaller margins of error.

In an *experiment*, a specific *treatment* is imposed on each subject in order to observe a response. A good experiment is the only way to convincingly establish that a stimulus actually *causes* a certain response. Important examples of experiments are:

Clinical trials of new drugs or medical treatments,
Industrial research to improve products,
Scientific research to understand new subjects.

The Physician's Health study is an example of a clinical trial.

A simple experimental design,

Treatment → Observations

often fails because uncontrolled influences are *confounded* with the effect of the treatment. The *placebo effect* is a source of confounding in clinical trials.

A *randomized comparative experiment* uses two or more groups of subjects to compare two or more treatments. Outside influences act on all the groups, so their effect is common to all outcomes. Subjects are assigned to groups at random. This avoids bias and allows probability to help interpret the results.

Random Allocation ↗ Group 1 → Treatment → Observation
Random Allocation ↘ Group 2 → Control → Observation

Experimental treatments can be combinations of two or more *factors*, such as pressure and temperature in a study of an industrial process. The experiment then reveals *interactions* (joint effects) of these factors as well as the effect of each separately.

More complex experimental designs use randomization also, but not the all-at-once random selection employed in simple designs. Sometimes subjects are divided into homogeneous *blocks* (such as men and women), then randomly assigned to treatments separately within each block. *Latin square designs* are used when there are three factors, each with the same number of different values.

Skill Objectives

1. Identify *population* and *sample* in a sampling or experimental situation.
2. Detect clear sources of *bias* in a sampling situation.
3. Use the table of random digits to select a *random sample* from a small population.
4. Recognize *confounding* of the effects of two variables.
5. Outline in a diagram the design of a *randomized comparative experiment* to compare the effects of several treatments.
6. Use the *table of random digits* (or coin tossing) to carry out the randomization required in an experimental design.
7. Be able to lay out a 3x3, 4x4, or 5x5 *Latin square* and randomly assign levels of an experimental factor and two blocking factors.

Self-test

MULTIPLE CHOICE

1. A group of medical students were given salt pills, but were told they were taking a strong stimulant. After taking the pills, the students talked excitedly and otherwise acted high. This is an example of
   a. the double blind technique.
   b. the placebo effect.
   c. a completely randomized experiment.
   d. a social experiment.

2. A public opinion firm wants information about the attitude of owners of common stock toward the Reagan Administration's economic policy. They interview 700 persons who hold accounts at Merrill Lynch, the largest retail stockbroker. In statistical language, these 700 people are
   a. the population.
   b. a sample.
   c. a statistic.
   d. a parameter.

3. Because a certain type of cancer behaves differently in men and women, a clinical trial to compare several treatments is designed as follows. First, the patients are grouped by gender, and then the men and women are separately assigned at random to the three treatments. This is an example of
   a. a Latin square design.
   b. a simple random sample.
   c. a design with two blocks.
   d. a multistage sample.

4. A study shows that students who took Latin in high school have much higher scores in a test of verbal skills than those who took no Latin, and concludes that Latin improves verbal skills in English. But these students who chose Latin are probably already unusually bright and interested in language, so that the conclusion is not valid. This is an example of
   a. an experiment whose results don't generalize.
   b. an experiment with observed control.
   c. the placebo effect.
   d. confounding variable.

5. Purdue University plans to survey the education goals of 200 Indiana high school freshmen. A statistician suggests increasing the sample size to 500 in order to
   a. reduce the bias of the result.
   b. increase the bias of the result.
   c. reduce the margin of error of the result.
   d. increase the margin of error of the result.

6. Random allocation in an experiment is used to avoid
   a. lack of precision.
   b. bias.
   c. confounding.
   d. double blinds.

7. The basic ideas of statistically designed experiments are
   a. bias and precision.
   b. randomization and blocking.
   c. randomization, double-blind, and bias.
   d. randomization and control.

8. A researcher interested in male attitudes toward women in professional occupations administers a questionnaire to 100 randomly selected male undergraduates at her college. She announces that "Most American males now accept women as lawyers, professors, and in other high status occupations." This conclusion may be invalid because
   a. there was no control group.
   b. fraternity and non-fraternity men have different attitudes toward women.
   c. many men show bias toward women.
   d. male undergraduates are not representative of all American males.

9. Monthly information about employment and unemployment in the United States is gathered by
   a. the Gallup Poll.
   b. the Current Population Survey.
   c. the decennial census.
   d. the national survey of manufacturers.

10. Which of the following is true of a table of random digits?
    a. The first digit in any row has chance 1/10 of being a 1.
    b. The first pair of digits in any row has chance 1/100 of being 11.
    c. Each row of 40 digits contains exactly 4 digits which are 1.
    d. Both a and b.
    e. All of a, b, and c.

11. After a Presidential address on relations with Russia, a public opinion firm wanted to survey the reactions of American adults. They called 600 persons chosen at random from telephone directories. The *population* in this sampling survey is
    a. all American adults.
    b. all Americans with telephones.
    c. all Americans listed in telephone directories.
    d. the 600 persons who were called.

12. The precision of a Gallup Poll probability sample of 1500 persons from the population of 165 million American residents age 18 and over can be expressed by saying "95% of all such samples will give a result falling within ±3 percentage points of the true population value."

    If the sample size were increased to 4000 persons, you would expect the ±3 point margin of error to change to
    a. ±4 points.
    b. ±3 points (no change).
    c. ±2 points.
    d. ±0 points (perfect accuracy).

13. The evidence that smoking causes lung cancer, while strong, is not completely convincing because
    a. lung cancer rates among smokers are only a little higher than among nonsmokers.
    b. randomized controlled experiments to compare smokers and nonsmokers cannot be done.
    c. lung cancer can also be caused by other factors such as breathing asbestos fibers.
    d. the margin of error in the samples of people studied is too large to allow a clear conclusion.

14. Medical experiments, for example, to compare a new vaccine with a placebo, are often double blind. This means that
    a. subjects are assigned randomly, or blindly.
    b. the subjects are not told which treatment they received.
    c. the placebo group is used for comparison.
    d. neither the subject nor the diagnosing physician knows which treatment the subject received.

15. The purpose of randomization in assigning units to treatments in an experiment is
    a. to create roughly similar groups on each treatment.
    b. to make it difficult to predict the results beforehand.
    c. to avoid the placebo effect.
    d. to avoid deception.

ANSWERS

1. (b), 2. (b), 3. (c), 4. (d), 5. (c), 6. (b), 7. (d), 8. (d), 9. (b), 10. (d), 11. (a), 12. (c), 13. (b), 14. (d), 15. (a).

Sample problems

16. Use the random digits to choose an SRS of five of the following Statistics students to be given special tutorial sessions. Show your work in detail.

00 Austin	06 Conway	12 Hallman	18 Rouse
01 Axsom	07 Curry	13 Iunghuhn	19 Sears
02 Battreall	08 Davis	14 Miller	20 Sowers
03 Baxter	09 Durand	15 Piper	21 Thrasher
04 Blevins	10 Fancher	16 Quigley	22 Wicker
05 Burns	11 Granack	17 Roth	23 Young

RANDOM DIGITS

24013 22498 03316 32337 79367 16856 41267
00694 05977 19664 02150 43163 12724 04178

17. An experiment is planned to test whether an ingredient from marijuana will relieve the nausea often caused by anticancer drugs. A group of 20 cancer patients who are suffering such nausea is available. Their names are

Abbott	Edwards	Horowitz	Moore
Ardmore	Freund	Johns	Protter
Bozivich	Gans	Klipsch	Robards
Cabell	Grutzner	Lindley	Shilling
Denning	Hewlitt	McMahan	Winograd

Outline the design of an experiment to test the effect of the marijuana ingredient on nausea. Briefly discuss what precautions other than your basic design should be taken to ensure valid results.

18. A sociologist wanted to investigate the attitude of women living in San Diego, California, toward working mothers of pre-school children. She took a random sample of San Diego residential addresses and sent students to interview any adult woman living at the addresses chosen. The students worked only between one and four p.m. on weekdays. The women interviewed had somewhat negative attitudes toward working mothers.

    Explain *why* the sample result is probably *biased,* and how the sample result is probably systematically different from the truth about the population.

ANSWERS

16. Step 1: Label the students (2 digits each) as shown above.

    Step 2: Read random digits in two-digit groups. Ignore groups not used as labels or repeated. SRS is

    01 = Axsom
    03 = Baxter
    23 = Young
    06 = Conway
    05 = Burns

17. A diagram presents the design most clearly

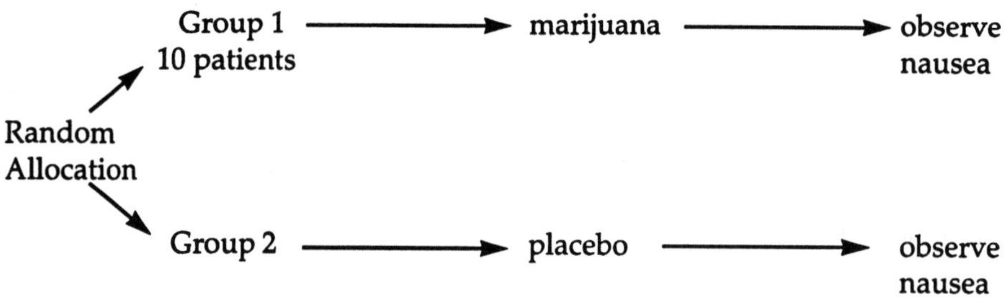

Note that some treatments must be given to group 2 to avoid the placebo effect. Since degree of nausea is a bit subjective, the experiment should be *double-blind.* Both groups must be treated identically except for the content of the pills they take.

18. Working women tend not to be at home between one and four p.m. on weekdays. They will be underrepresented in the respondents because of this. This causes bias.

    Working women are probably more in favor of working mothers than are women who choose to be at home. So sample attitudes will be more negative than the true attitudes of the population of women.

SHOW TWO: PICTURE THIS: ORGANIZING DATA

Data are summarized and made easier to grasp by *graphical methods* and numerical *descriptive statistics*. *Exploratory data analysis* is the art of looking for unanticipated patterns in data, as opposed to formal statistical inference, which seeks to answer specific questions posed in advance.

The distribution of values of a single variable is described by a *histogram* showing the *frequencies* of each value. When examining a histogram, look first for the overall pattern of the data and then for *outliers*. The general principle is

Data = Smooth + Rough

Descriptive statistics for a single variable summarize its location (*median* or *mean*) and its spread or variability (*range, quartiles*). The *five-number summary* quickly describes both location and spread, and is pictured in a *boxplot*.

The relationship between two variables is described by a *scatterplot*. If the relationship is linear, a *regression line* can be fitted to the scatterplot and is used to predict one variable from a given value of the other. In this case, the line represents the overall pattern (the smooth) and the scatter of points about the line is the rough.

Scatterplots can also show the relationship between three variables. These plots are hard to see clearly unless color or motion is used. Computer graphics makes it possible to do this.

Skill Objectives

1. Construct a *histogram* of small ($n \leq 100$) data sets and recognize outliers.
2. Compute the *mean* and *median* of small data sets.
3. Compute the *extremes, quartiles* (and thus the five-number summary), and *range* of small data sets.
4. Draw a *box plot* based on a five-number summary.
5. Draw a *scatterplot* of bivariate data and *fit a line* by eye when the scatterplot shows a clear linear line.
6. Be able to *predict* the response Y to a given X, both graphically using a fitted line, and arithmetically when given the equation of a regression line.

Self test

MULTIPLE CHOICE

1. The third quartile of scores on a statistics exam was 83. This means that
   a. one quarter of the scores were above 83.
   b. one half of the scores were above 83.
   c. three quarters of the scores were above 83.
   d. 83 was three standard deviations above the mean.

2. Over the past ten games, the mean number of points scored by a certain basketball player is 12 points per game. We may correctly conclude that
   a. half of the games he scored more than 12 points, and half of the games he scored less than 12 points.
   b. he mostly typically scored 12 points a game.
   c. he scored a total of 120 points over the past ten games.
   d. all of the above.

3. The median of the numbers 1, 5, 4, 4, 8, 11, and 9 is
   a. 4.
   b. 5.
   c. 6.
   d. 8.

4. You measure the age, marital status, and earned income of a random sample of 1289 women. The number of variables that you have measured is
   a. three - age, marital status, and income.
   b. two - age and income. Marital status doesn't count because it doesn't have a unit like years or dollars.
   c. four - age, marital status, income, and number of women.
   d. 1289 - the size of the sample.

5. Even when you compute numerical descriptive statistics, a graph often helps present your conclusion. For example, you would often draw a *boxplot* to accompany
   a. a correlation coefficient.
   b. a mean and standard deviation.
   c. a five-number summary.
   d. a bivariate frequency table.

6. You would draw a *scatterplot* to
   a. show how the number of television sets in the U.S. has increased over time.
   b. compare the number of television sets in the U.S., England, Japan, and Russia.
   c. display the five-number summary of the numbers of television sets in the 50 states.
   d. show the relationship between the number of television sets and the number of television stations in the 50 states.

7. The median is a measure of
   a. the center or location of a set of data.
   b. the spread or dispersion of a set of data.
   c. the most common value in a set of data.
   d. the association in a set of data.

8. If your exam score is at the 70th percentile of the scores for the entire course, you scored
   a. above the median.
   b. right at the median.
   c. below the median.
   d. can't tell from the information given.

9. Five people are asked how many pennies they have in their pockets. The answers are 0, 1, 1, 5, and 3. The mean number of pennies in the pockets of the five people is
   a. 2.
   b. 10.
   c. 5.
   d. 1.

10. You are drawing a scatterplot to show the relationship between the hours each student spent studying and their scores on an examination. You decide that
    a. exam score belongs on the horizontal axis.
    b. hours studied belong on the horizontal axis.
    c. either variable can go on the horizontal axis.
    d. a scatterplot is not appropriate in this case.

ANSWERS
1. (a), 2. (c), 3. (b), 4. (a), 5. (c), 6. (d), 7. (a), 8. (a), 9. (a), 10. (b).

Sample Problems

11. Here are data on the rate of price increases in a number of countries. The numbers are the percentage increase in consumer prices from 1979 to 1980. (Source: 1984 *Statistical Abstract of the U.S.*, Table 1517.)

Argentina	100.8	Iran	20.7
Australia	10.2	Israel	131.0
Brazil	82.8	Japan	8.0
Canada	10.2	Mexico	26.4
Chile	35.1	Nigeria	11.4
Denmark	12.3	Peru	59.2
Egypt	20.7	Philippines	17.8
France	13.3	Spain	15.6
Germany	5.5	Switzerland	4.0
Greece	24.9	United States	13.5
India	11.4		

a. Compute the five-number summary of this set of data (do not use deciles).

b. Make a boxplot from you five-number summary. Are there any outliers? If so, which countries are they?

12. An agricultural researcher is studying the effect of planting rate (number of plants per acre) on corn yields. He carries out an experiment by planting at different rates of different plots of ground (chosen at random) and measuring the yield. Here are the data from 12 plots.

Planting Rate (thousands)	Yield (bushels/acre)
12	130.5
12	129.6
16	142.5
16	140.3
16	143.4
20	145.1
20	144.8
20	144.1
24	147.8
24	148.4
28	134.8
28	135.1

a. Make a scatterplot of the data.

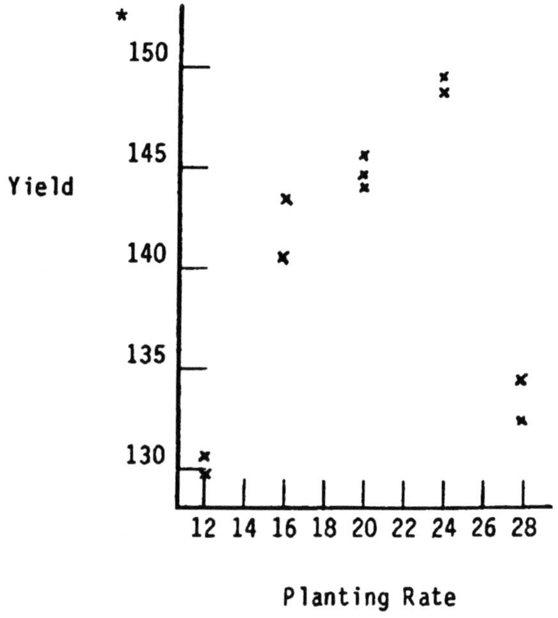

b. Describe in words the overall pattern of the relation between planting rate and yield. Are there any outliers in the data? Would you be willing to use a straight line fitted by a least squares regression program to predict corn yield on a plot with 22,000 plants per acre?

13. A study of hospital patients measured the number of days each was hospitalized (x) and the amount (y, in hundreds of dollars) of the hospital bill. A scatterplot showed good fit to a straight line. A regression program produced the regression line.

$$y = .50 + 4.3x$$

Use this equation to predict the bill of a patient who spends four days in the hospital.

ANSWERS

11. a. First arrange in increasing order:

4.0 5.5 8.0 10.2 10.2 11.4 11.4 12.3 13.3 13.5 15.6 17.8 20.7 20.7 24.9 26.4 35.1 59.2 82.8 100.8 131.0

There are 21 countries, so median is 11th (15.6), and quartiles are the medians of the 10 observations on each side of 15.6.

First quartile = 10.2 + 11.4/2 = 21.6/2 = 10.8

Third quartile = 26.4 + 35.1/2 = 30.75

b. The five-number summary is 4.0 10.8 15.6 30.75 131.0

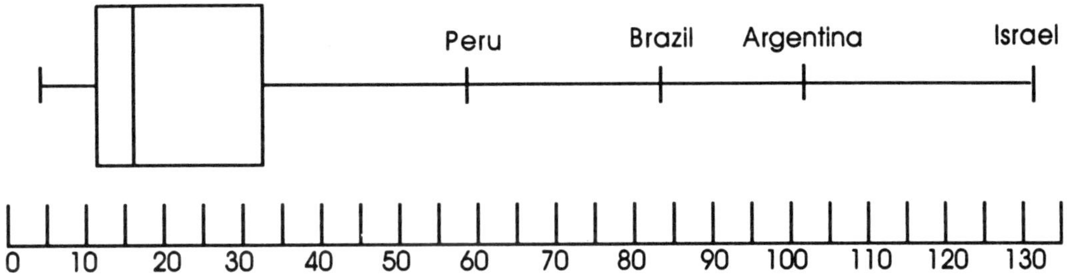

The distribution is very skewed to the right. The four countries shown above could be called outliers. This is a matter of judgement.

12. Yield increases with planting rate until about 24,000 plants per acre. Beyond that rate, yield decreases. Seen this way, there are no outliers.

    A straight line does not describe the pattern - the relation is curved. I would not use a regression line for prediction.

13. $y = .50 + (4.3)(4)$
    $= .50 + 17.2$
    $= 17.7$

    Since y is in *hundreds* of dollars, the answer is

    $y = \$1,770.00$

SHOW THREE: PLACE YOUR BETS: PROBABILITY

Random phenomena have unpredictable outcomes that nonetheless follow a predictable pattern in many repetitions. Examples of phenomena include:

Games of chance (rolling dice, roulette wheels),
Heredity (hair or eye color),
Random sampling and experimental randomization.

The long term pattern of random outcomes is described by a *probability* model. This consists of a *sample space* or list of possible outcomes *S*, and a probability or long run proportion *P*(s) assigned to each outcome *s* in *S*. Any assignment of probabilities satisfying two simple laws is legitimate. Which assignment is actually correct can only be determined by observing many outcomes.

A statistic is a number computed from a sample. When the sample is random, the probabilities with which the statistic takes its possible values make up the *sampling distribution* of the statistic.

When the statistic is an average or a percentage in a large sample, the sampling distribution is close to a *normal curve*. Normal curves are symmetric, so the mean and median are equal. This is not true in a *skewed* distribution. A particular normal curve is completely described by its location (measured by the mean) and spread (measured by the *standard deviation*). The *68-95-99.7* rule describes how probability is distributed in a normal distribution.

The *central limit theorem* says that averages over large samples are always nearly normally distributed, and that the standard deviation of the distribution decreases with *n*, the square root of the sample size. So statistics from large samples are less variable than the same statistics from a smaller sample.

If we know the probabilities of the possible outcomes of a random phenomenon, we can compute the *expected value*. The *law of large numbers* says that the average of many actual outcomes must eventually be close to the expected values. If the outcomes have high variability (as do the winnings in most games of chance), a very large number of repetitions may be needed for the average to be close to the expected value. The *n* in the central limit theorem describes how variability decreases as the number of repetitions, *n*, increases.

Skill Objectives

1. Describe the *sample space* of a random phenomenon.
2. Identify legitimate and illegitimate *probability models* by checking Laws 1 and 2.
3. Be able to *estimate probabilities* as long run proportions and construct the *sampling distribution* of simple statistics repeated trails.
4. Compute *probabilities* of events by counting outcomes when all outcomes are equally likely.
5. Locate the *mean* and *standard deviation* from a graph of a normal curve.
6. Use the *68-95-99.7* rule to compute normal probabilities.

7. Compute the *expected value* of a numerical random outcome when given the probability of each value.
8. Use the *central limit theorem* together with skill 6 above to make probability statements about any number of trials when the standard deviation for some number *n* of trials is given.

Self-test

MULTIPLE CHOICE

1. In a certain lottery there is a probability of .1 of winning $100.00, and a probability of .9 of winning nothing. The lottery costs nothing to enter. Your expected winnings are therefore
   a. $100.00.
   b. $0.00.
   c. $10.00.
   d. none of the above.

2. Referring to **Problem** 1, the probability of .1 means
   a. every tenth person will win.
   b. in ten people, there will be one winner.
   c. about 10% of those who enter will win, if a large number enter.
   d. none of the above.

3. In any probability model, which of the following are true?
   a. All probabilities are equal.
   b. All probabilities are random.
   c. All probabilities are between 0 and 1.
   d. All probabilities are zero.

4. Scores on an American College Testing aptitude test are approximately normally distributed with mean 18 and standard deviation 6. The middle 95% of scores fall between approximately
   a. 12 and 24.
   b. 6 and 30.
   c. 0 and 36.
   d. Cannot tell from the information given.

5. In 1977, the 88 quarterbacks on NFL rosters earned an average salary of $58,750.00 or $89,354.00, depending on whether you report the mean salary or the median salary. Which of these numbers is the mean salary?

a. $58,750.00.
b. $89,354.00.
c. Cannot tell--either is equally plausible.

6. A normal curve is uniquely described by
   a. the mean and standard deviation.
   b. a scatterplot.
   c. the median and range.
   d. the central limit theorem.

7. College Board (SAT) scores for a reference population are approximately normally distributed with mean 500 and standard deviation 100. Approximately what percentage of the population have SAT scores above 500?
   a. 16%
   b. 34%
   c. 50%
   d. 68%

8. If the probability of snow tomorrow is 0.7, what is the probability of no snow tomorrow?
   a. 0.7
   b. 0.3
   c. 0.0
   d. Cannot tell--could be any number between 0 and 1.

9. You want to find the average speed of vehicles on the interstate highway you are driving on. So you adjust your speed until the number of vehicles you are passing equals the number passing you. You have found
   a. the mean speed.
   b. the median speed.
   c. the mode speed.
   d. none of these.

10. In order to determine the percentage of adult residents of Denver, Colorado, who ski, a random sample of 750 residents are interviewed by telephone. Of those interviewed, 74% said they were skiers. The number 74% here is
    a. a sample.
    b. a parameter.
    c. the population.
    d. a statistic.

ANSWERS

1. (c), 2. (c), 3. (c), 4. (b), 5. (b), 6. (a), 7. (c), 8. (b), 9. (b), 10. (d).

Sample Problems

11. Above the axis printed below, draw a picture of a normal curve with mean 36 and standard deviation 2.

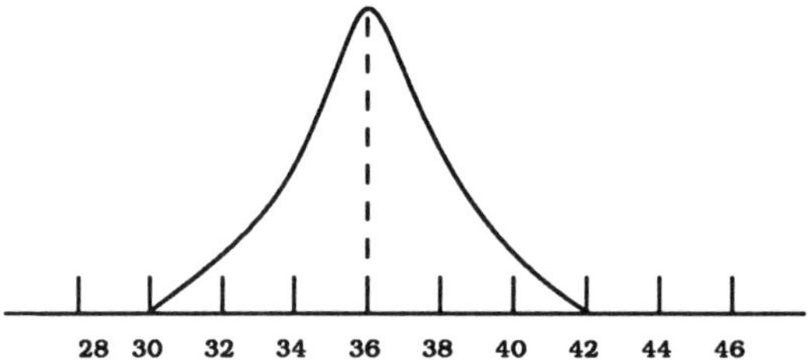

12. A couple decides to have three children. Of course, each child will be either a girl or boy.

    a. Using G for girl and B for boy, list all the possible sequences of sexes of their children in order of birth.

    b. All outcomes in your sample space of part **a.** are equally likely. What is the probability of each outcome?

    c. What is the probability that the family will have exactly one girl among its three children?

13. We examine a large number of families having exactly three children each. If we choose 100 such families at random, the average number of girls per family will vary from sample to sample according to a normal distribution with mean 1.5 and standard deviation 0.087.

    a. Between what values will the average number of girls fall in 95% of all such samples of size 100?

    b. If we choose 1000 families at random, instead of 100, what sampling distribution will the average number of girls have?

    c. Between what values will the average number of girls fall in 95% of all random samples of size 1000?

ANSWERS

11. Curve should be symmetric about 36 total width about ±6 or 30 to 42 change of curvature at 34 and 38.

12.

a.

GGG	BGG
GGB	BGB
GBG	BBG
GBB	BBB

b. There are 8 outcomes, so each probability is 1/8.

c. This even consists of the outcomes

GBB BGB BBG,

so it has the probability 3/8.

13. a. By the *68-95-99.7 rule,* 95% fall in mean ±2 standard deviation
1.5 ± (2)(.087)
1.5 ± .174
or 1.326 to 1.674.

b. Normal, mean still 1.5, standard deviation (because sample of 1000 is 10 times 100)

$$\frac{.087}{\sqrt{10}} = .0275$$

c. By the 68-95-99.7 rule again mean ±2 standard deviation
1.5 ± (2)(.0275)
1.5 ± .055
or 1.445 to 1.555.

As expected, the average over 1000 families is less variable than an average over 100 families.

SHOW FOUR: CONFIDENT CONCLUSIONS: STATISTICAL INFERENCE

Formal statistical inference, as opposed to exploratory data analysis, is based on calculations of probability. We use methods that have known probability of making correct decisions. Usually the probability of a correct decision is chosen to be quite high, often .95.

Inference is based on the sampling distribution of a sample statistic. We consider two statistics: the sample *proportion,* $\hat{p}$, and the sample mean $\bar{x}$. The central limit theorem says that the sampling distributions of these statistics are close to normal when the samples are large.

A number that describes a population is called a *parameter*. Suppose that the population proportion of some characteristics is the unknown parameter p. We draw a simple random sample of size n and compute the sample proportion $\hat{p}$. The sampling distribution of $\hat{p}$ is approximately normal with mean equal to p and standard deviation equal to

$$\sqrt{\frac{p(100 - p)}{n}} \; .$$

(Both p and $\hat{p}$ are measured in percent in these formulas.) Since p is not known, we replace p by $\hat{p}$ in the last formula in order to estimate the standard deviation of $\hat{p}$.

A *confidence interval* estimates an unknown parameter by computing an interval from a sample by a method that has known probability (called the *confidence level*) of producing an interval that contains the true value of the parameter.

A *95% confidence interval for a population proportion* p is obtained as follows. Take a simple random sample of size n from the population. Compute the sample proportion $\hat{p}$. Then in 95% of all samples, the known p falls in the interval

$$\hat{p} \pm 2\sqrt{\frac{\hat{p}(100 - \hat{p})}{n}} \; .$$

In nontechnical writing, this summarized by giving the sample value $\hat{p}$, the confidence level (95%) and the numerical value of the *margin of error*

$$\pm 2\sqrt{\frac{\hat{p}(100 - \hat{p})}{n}} \; .$$

Confidence higher than 95% can be obtained, but at the price of a larger margin of error. Note that the margin of error decreases at the rate $\sqrt{n}$ as the sample size n increases. These properties are common to many kinds of confidence intervals.

We can also give confidence intervals for the unknown mean μ of a population. Suppose that the standard deviation of a population is known to have the value σ. This happens in quality control settings, where σ measures the variability of a production process. The variability is often known, but the mean μ may change and must be controlled at some target μ value. If we take a simple random sample of size n and compute the sample mean $\bar{x}$, the sampling distribution of $\bar{x}$ is approximately normal with mean μ and standard deviation $\sigma / \sqrt{n}$.

A *95% confidence interval for a population mean* is given by

$$\bar{x} \pm \frac{2s}{\sqrt{n}}.$$

In a control chart, the same information is arranged differently. Plot the means x of successive samples vertically against time horizontally. Draw on this chart a *center line* at the target value of μ. Then draw *control limits* at

$$\mu \pm \frac{2s}{\sqrt{n}}.$$

Only 5% of all samples will have $\bar{x}$ outside the control limits when the population mean is truly equal to μ. So a point plotted *out of control* is evidence that the process mean has moved away from its target.

The purpose of a control chart is not to catch individual defective products, but to keep the production process properly adjusted. It combines statistical inference with a graph that shows the state of the process.

Hidden variables that influence the variables measured can make both exploratory analysis and formal inference misleading. for example, the direction of an apparent relationship between two variables can even be reversed when a hidden variable is taken into account. Statistical methods must be used thoughtfully, not methodically.

Skill Objectives

1. Identify *parameter* and *statistic* in simple inferential settings.
2. State the *mean* and *standard deviation* of the sampling distribution of $\hat{p}$ and $\bar{x}$ when given the population parameters, and use the *68-95-99.7 rule* to make probability calculations for $\hat{p}$ and $\bar{x}$.
3. Give *95% confidence intervals* for p and μ (from large samples), and recognize when each is appropriate.
4. Draw an *x control chart* for repeated samples from a process with known σ and stated target μ for the mean.
5. Be able to *aggregate* a three-way table into a two-way table by summing over one variable.
6. Be able to *describe association* in a two-way table by computing appropriate percents, and to recognize how the direction of association can reverse due to aggregation.

Self-test

MULTIPLE CHOICE

1. Statistics may be defined as
   a. the study of record-keeping for sports or governments.
   b. the study of methods for collecting data, describing data, and drawing conclusions from data.
   c. the study of the properties of numbers and of the solutions of numerical equations.
   d. the study of the natural world by experiments.

2. In statistical terminology, a parameter is
   a. a number that is computed from sample data.
   b. the margin of error in a confidence interval.
   c. a number that describes the population.
   d. a fraction or proportion.

3. A control chart is a plot whose purpose is
   a. to illustrate the 68-95-99.7 rule.
   b. display the relationship between two variables.
   c. display a five-number summary of a set of quality control data.
   d. to show whether or not a production process is operating at its target setting.

4. Increasing the size of a simple random sample has the effect of
   a. decreasing the margin of error of a 95% confidence interval.
   b. increasing the margin of error of a 95% confidence interval.
   c. increasing the confidence level of a 95% confidence interval.
   d. decreasing the confidence level of a 95% confidence interval.

5. Suppose that, in fact, 61% of all U.S. adults favor the President's tax reform program. A Gallup Poll of 1500 adults finds that 59% of the sample favor the program. In this example, the number 61% is
   a. a statistic.
   b. a parameter.
   c. a sample.
   d. a population.

6. The diameter of bearings produced by a company varied due to small variations in the material and in the production process. The distribution of bearing diameters is normal with mean 1.000 cm. and standard deviation 0.002 cm. So, 95% of all the bearings produced have diameters between

a. .998 and 1.002 cm.
b. .996 and 1.004 cm.
c. .994 and 1.006 cm.
d. .98 and 1.02 cm.

7. The company in Question 6 takes regular quality control samples of $n = 4$ bearings, and measures their diameters. In 95% of all samples, the sample mean diameter $\overline{x}$ falls between
   a. .997 and 1.003 cm.
   b. .996 and 1.004 cm.
   c. .999 and 1.001 cm.
   d. .998 and 1.002 cm.

8. You have calculated 95% confidence interval with a margin of error of ±3. Since that margin of error is too large, you double the size of your sample. Now the margin of error will be
   a. ±1.50.
   b. ±2.12.
   c. ±4.24.
   d. ±6.00.

9. Suppose that, in fact, 74% of all college freshmen would agree that "being well-off financially" is one of their important goals. A sociologist asks this question of a random sample of 1500 freshmen and computes the percent $\hat{p}$ who agree. The mean of the sampling distribution of $\hat{p}$
   a. is equal to 74%.
   b. varies according to a normal distribution.
   c. falls within ±3% of 74%.
   d. both b and c.

10. Formal statistical inference differs from exploratory data analysis (EDA) in that
    a. inference uses numerical statistics, while EDA uses only graphs.
    b. hidden variables can make EDA misleading, but do not deceive formal inference.
    c. formal inference cannot be combined with graphs of the data, but EDA can.
    d. formal inference uses probability to answer specific questions, while EDA looks for any unusual pattern in the data.

ANSWERS
1. (b), 2. (c), 3. (d), 4. (a), 5. (b), 6. (b), 7. (d), 8. (b), 9. (a), 10. (d).

Sample Problems

11. A bistable storage CRT has a set of fine (500 line pairs per inch) mesh screens behind the viewing surface. The mesh is stretched and welded onto a metal frame during assembly. Too little tension at this stage will cause wrinkles while too much will tear the mesh. High rejection rates (20% or more) are being experienced. An engineer develops a tension measuring device (the output is in millivolts, with higher voltage showing lower tension) that should allow better control of the tension/welding system. A careful study shows that the process standard deviation is $\sigma = 43$ mv. when operating properly, and that the target value for the mean should be $\mu = 275$ mv.

    Here are data from 20 successive samples of size $n = 4$ from this process. Make an $\bar{x}$ control chart, including the center line and control limits. Mark any out-of-control points and comment on your findings.

Sample	1	2	3	4	5	6	7	8	9	10	11	12
$\bar{x}$	253	269	262	254	259	252	268	286	266	282	282	285

13	14	15	16	17	18	19	20
292	304	351	299	315	314	318	266

12. You are curious about the common practice of using mayonnaise jars for home canning of vegetables. These jars are the proper size, but the glass is thin and may crack during processing. You can 200 quarts of tomatoes in such jars. Of these, 14 crack. Give a 95% confidence interval for the percent of all mayonnaise jars that would crack if used for canning.

13. Here is a table of advanced degrees for 1981 in two subjects, broken down by sex and level. (Source: 1984 *Statistical Abstract of the U.S.*, Table 276.)

	Languages			Mathematics	
	Male	Female		Male	Female
Master's	694	1410	Master's	1692	875
Doctorate	274	314	Doctorate	614	114
Total	968	1724	Total	2306	989

a. What percentage of Master's degrees in languages were earned by women?

b. What percentage of Master's degrees in languages and mathematics combined were earned by women?

ANSWERS

11. Center line = 275

$$\text{Control limits} = \mu \pm 2\frac{\sigma}{\sqrt{n}} = 275 \pm 2\frac{43}{\sqrt{4}} = 232 \text{ and } 318$$

There is an upward drift, with **Sample #15** out of control and three others very close to the limit. The tension must be readjusted.

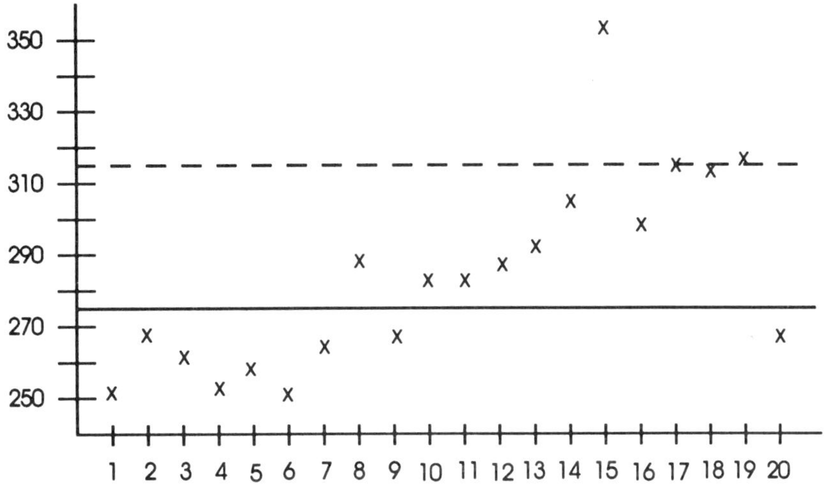

12. The sample proportion of cracked jars is

$$\hat{p} = 14/200 = 0.07 = 7\%$$

So, an approximate 95% confidence interval for the proportion p of all such jars that would crack is

$$\hat{p} \pm 2\sqrt{\frac{\hat{p}(100 - \hat{p})}{\hat{n}}}$$

$$7\% \pm 2\sqrt{\frac{(7)(93)}{200}}$$

$$7\% \pm (2)\sqrt{3.255}$$

$$7\% \pm 3.61\% \text{ or } 3.39\% \text{ to } 10.61\%$$

13. a.

$$\frac{1410}{694 + 1410} = \frac{1410}{2140} = 0.67 = 67\%$$

b.

$$\frac{1410 + 875}{694 + 1410 + 1692 + 875} = \frac{2285}{4671}$$

$= 0.489 = 49\%$

[The aggregated data show women earning very close to 1/2 of Master's degrees. But this hides the fact that they earn 2/3 of all Master's degrees in language, but only 1/3 (34%) in mathematics.]

Part III Social Choice
William F. Lucas
Claremont Graduate School

INTRODUCTION

Mathematics has traditionally studied the nature and algebra of numbers, the patterns and structure of space, and the continuous mathematics of calculus with its great power to describe the intricate workings of the physical universe. Many of the major advances of mathematics in the twentieth century have been concerned with more human affairs, the interface of humans with machines and large systems, and the highly discrete mathematics so compatible with modern *digital* computers. The Social Choice television programs and the accompanying text Chapters 9-12 are intended to provide a glimpse into how a few of the new mathematical subjects relate to human activities: their behavior, organizations, interactions, institutions, and decision making. This should also provide the viewers and readers with some actual assistance in arriving at their own personal decisions, as well as provide some appreciation for recent mathematical achievements.

Surprisingly few specialized mathematical prerequisites are needed to comprehend this material. Basic ideas from arithmetic and elementary set theory are used. A mild dose of the illusive notion called "mathematical maturity" will prove useful at times. The concept of an expected value (average) and the ability to solve two linear algebraic equations in two unknowns is required to appreciate and compute optimal mixed strategies for the two-person, zero-sum (matrix) games. The main tool used throughout is that of a table (called a *matrix* by mathematicians). It is used to list one's available options, the potential outcomes of a decision, or one's individual preferences. The use of the tables (spread sheets) and graphs (road maps) are crucial to much of modern discrete mathematics.

Note that the third and fourth teaching shows on game theory correspond to only *one* chapter in the textbook. An additional chapter in the text on fair division and apportionment concentrates on topics that were spread rather lightly over the videos.

The Review Vocabulary at the end of each chapter in the text provides a good indication of the main ideas to be learned. To fully appreciate these concepts and to be able to incorporate the techniques into one's own personal decision making, one will need to practice by solving many of the exercises presented in the textbook.

OVERVIEW SHOW

Arriving at decisions is a common human activity. Individuals, each endowed with free will and different desires, frequently select from alternate courses of action in an attempt to improve their quality of life. They strive to reach certain goals, but their choices are complicated by an uncertain future, competing interest, tradeoffs between different objectives, and the unpredictable actions of others. Modern mathematics has come to the aid of contemporary decision makers. A score of new subjects have been introduced in recent decades to quantify and assist in the different aspects of a decision. A few of these new approaches are described in the subsequent programs.

A decision problem consists of individuals with options, the potential outcomes resulting from these choices, and one's relative preferences (or utility) for these outcomes (or payoffs). *Decision Theory* is concerned with selecting a course of action in light of an uncertain future. Should one carry his or her umbrella on a day when there is a fifty-fifty forecast for rain? The answer, of course, depends upon what benefits or penalties one associates with lugging around an umbrella on a sunny day, getting caught unprepared in the rain, etc. Such decisions can be viewed as a "game against nature," although we do not normally assume that nature is aiming to defeat us.

Game Theory is a mathematical approach to studying conflict and cooperation, where one's competitors are other persons who are also out to maximize their own gains. Some games are strictly competitive (the *zero-sum* games), where one person's gain equals the protagonist's loss. Many parlor games, such as chess, or team sports like basketball, are viewed as *two-person*, zero-sum games. The mathematical theory attempts to derive optimal strategies for playing such contests. Most encounters are, however, not zero-sum in nature and typically involve aspects of both competition and cooperation. The best known of all two-person, *general-sum* games is known as the Prisoner's Dilemma or as the Tragedy of the Commons. It arises when the pursuit of individual goals results in the worst available global outcome for the contestants. This is illustrated by arms races, trade wars, excessive pollution, and the over exploitation of a common resource, such as over fishing to the point of near extinction. If everyone else receives a vaccination against a contagious disease like whooping cough (pertussis), then it is safer for me or my child to avoid the shot and its potential side effects. Regulation or trust are needed in order to void the social paradox present in the prisoner's dilemma or the "vaccination game."

Social Choice Theory is the mathematical subject which studies how groups in a democratic setting should arrive at a decision. The nature and properties of group decision mechanisms such as various voting methods are topics of analysis. Determining what sport to play at a picnic or selecting the menu for a party at a Chinese restaurant are typical examples of common group decisions, as are electing government officials and passing new legislation. Kenneth Arrow's famous impossibility theorem proved that every voting method can exhibit undesirable, unnatural, or counterintuitive results in some applications. Furthermore, the outcome of an election can depend upon the ordering of items in an agenda, upon falsification of one's true preferences, or on other manipulative type behavior. In short, voting is also a strategic contest requiring skill at playing this game.

The *Theory of Fair Division* provides schemes by which a group of people can divide up some benefits or losses in such a way that everyone of the participants feels that he or she received a fair share of goods or tasks. Dividing a cake among children or an estate among heirs are interesting applications. Bidding and monetary side payments may be used as in the case of two people bidding for one object, like a frisbee.

The Overview Program provides simple examples from these four mathematical subjects which were created to assist human decisionmakers in their tasks.

SHOW ONE: THE IMPOSSIBLE DREAM: ELECTION THEORY

Social choice is a subject devoted to the study of how groups can arrive at a decision. Individuals typically have different preferences over the available alternatives, but the group as a whole must select a particular ranking of these competing alternatives. In practice, a variety of different voting methods have been employed to determine one winner, or else to rank order all of the contenders.

A few of the common voting schemes are illustrated in the video show or the text, and are referred to as follows:

Majority rule
Plurality wins
Plurality with runoff
Borda count
Condorcet winner
Sequential elimination of the most unpopular
Sequential pairwise elimination (like in a tournament)

An individual expresses his or her own desires by ranking all of the possible candidates from the most preferred to the least wanted. A table describing the individual preferences for the full group of electors is called a *preference schedule*. It provides the critical input data for any group decisions. A voting method then turns these individual preferences into a choice for the group. This is illustrated by people selecting a restaurant for dinner, by sports writers picking the top-rated basketball team, or by delegates voting for the primary platform issue. The main example in the video program and text shows how five different voting methods could in fact elect five different winners in a five-person race.

The main lesson is that the outcome of an election can depend upon the method of voting used, the order in which the items are voted on, and strategic (insincere) voting. Voting is a contest and is vulnerable to manipulation. There is no perfect voting scheme when there are more than two alternatives and two or more voters. Any method will at times demonstrate undesirable results.

Skill Objectives

1. Learn the notion of preference schedule and how to use it.
2. Understand how some particular voting schemes work.
3. Be able to determine the winner from some popular voting methods, given the voters' preferences.
4. See how different voting schemes, or the order of voting, can lead to different winners.
5. Appreciate the strategic aspects of voting.
6. Appreciate the nature of Arrow's Impossibility Theorem.
7. Understand the recently popular voting method called approval voting.

Self-test

MULTIPLE CHOICE

1. How many different ways can one rank three alternatives if ties are not allowed?
   a. 2
   b. 3
   c. 6
   d. 9

2. How many different ways can one rank two choices if ties are allowed?
   a. 1
   b. 2
   c. 3
   d. 4

3. Are the following statements true (T) or false (F)?
   (1) Every voting method leads to the same winner. T F
   (2) Majority rule is a good voting method when deciding between one of two alternatives. T F
   (3) The choice of the voting method may determine who wins in an election. T F
   (4) Approval voting will never illustrate paradoxical outcomes or undesirable results. T F
   (5) One should never vote for a less preferred outcome over a more preferred one. T F
   (6) The order in which motions are voted upon may effect the results of an election. T F
   (7) Voters can never benefit in an election by falsifying their true preferences (i.e., voting in an insincere fashion). T F
   (8) The law of transitivity states that if A is preferred over B, and B is preferred over C, then C is preferred over A. T F
   (9) Any voting method will always produce outcomes for the groups' ranking which satisfy the law of transitivity. T F
   (10) Arrow's Impossibility Theorem states that any voting method with more than three voters will demonstrate undesirable results for any possible preference schedule for the voters. T F

4. Given that nine voters have the following preference schedule for three candidates A, B, and C.

Number of Voters	4	3	2
First Choice	A	B	C
Second Choice	B	A	B
Third Choice	C	C	A

Which candidate wins if the election is held by the following voting methods?
i. Plurality
a. A b. B c. C d. a tie results
ii. Plurality with runoff.
a. A b. B c. C d. a tie results
iii. Borda count which scores 3 points for each first choice, 2 for a second, and 1 for a third.
a. A b. B c. C d. a tie results

5. Assume that the nine voters in **Problem 4** knew the full preference schedule given there.
i. If approval voting were used in **Problem 4,** would you expect that the four voters who prefer A over B over C would approve of:
a. just A b. A and B c. A, B, and C d. no one
ii. If approval voting were used in **Problem 4,** do you expect that the two voters who prefer C over B over A would approve of:
a. just C b. C and B c. C, B, and A d. no one
iii. Would you expect that the four voters who prefer A over B over C would vote sincerely when the voting method is:
i. Plurality. a. yes b. no
ii. Plurality with a runoff. a. yes b. no
iii. Borda count scoring 3, 2, and 1. a. yes b. no

ANSWERS

1. (c); 2. (c); 3. (1) F, (2) T, (3) T, (4) F, (5) F, (6) T, (7) F, (8) F, (9) F, (10) F; 4:. (i) (a), (ii) (b), (iii) (b); 5: (i) (a), (ii) (b), (iii): (i) (a), (ii) (b), (iii) (b).

Sample Problems

6. List all possible ways three voters can rank three alternatives A, B, and C when ties are allowed.

7. The fifteen members of a county board of representatives have the following preference schedule over three outcomes A, B, and C.

Number of Voters	2	4	1	4	0	4
First choice	A	A	B	B	C	C
Second choice	B	C	A	C	A	B
Third choice	C	B	C	A	B	A

Determine which alternative would win if the choice were made by means of the

following voting methods:
i. Plurality.
ii. Plurality with a runoff.
iii. The alternative with most last place votes is eliminated first and then a runoff is held between the other two.
iv. The alternative with the fewest first place votes is eliminated first and then a runoff is held between the other two.
v. A Borda count where a first choice counts 3 points, a second choice 1 point, and a third choice 0 points.

8. Given that seven voters denoted by 1, 2, 3, 4, 5, 6, and 7 voted by approval voting on three candidates A, B, and C as follows.

Voter	1	2	3	4	5	6	7
Approves of	A			A	A		
candidates:	B	B	B	B	C		B
			C	C			C

i. Which candidate wins and which one finishes last?
ii. If these seven voters instead ranked the three candidates as follows and used a Borda count, scoring 3, 2, and 1 points for a first, second, and third choice, respectively, then which candidate wins and loses?

Voter	1	2	3	4	5	6	7
First choice	A	B	B	C	C	A	C
Second choice	B	A	C	B	A	C	B
Third choice	C	C	A	A	B	B	A

iii. If voters 4 and 7 knew the preference schedule in (ii), explain how they might change the way they voted when using the approval voting method in (i).
iv. In what sense could "approval voting" be named instead "disapproval voting"?

9. Nine voters voting for three candidates A, B, and C have the following preference schedule and this information is known to the voters.

Number of Voters	4	3	2
First choice	A	B	C
Second choice	B	A	B
Third choice	C	C	A

These nine voters are considering the following two sequential pairwise elimination agenda:
Agenda I: A vote between A and B is held first and then the winner runs against C.
Agenda II: A vote between B and C is held first and then the winner opposes A.

i. Which candidate wins if the voters vote sincerely in Agenda I? In Agenda II?
ii. Explain whether the four voters who prefer A over B and B over C might benefit by voting insincerely in Agenda I or Agenda II.

ANSWERS

6. A A B B C C AB AC BC A B C ABC B C A C A B C B A BC AC AB C B C A B A

7 i. A wins by 6 to 5 for B and 4 for C.
ii. B beats A in runoff 9 to 6.
iii. A is eliminated with 8 last place votes and then C beats B by 8 to 7.
iv. C is eliminated with 4 first place votes and then b beats A by 9 to 6.
v. 21 points for B, 20 for C, and 19 for A.

8. i. B wins with 5 approval votes, C receives 4, and A is last with 3.
ii. C wins with 3x3 + 2x2 + 2x1 = 15 points,
B receives 2x3 + 3x2 + 2x1 = 14 and,
A loses with 2x3 + 2x2 + 3x1 = 13.
iii. If voters 4 and 7 did not cast an approval vote for their second choice B, then their first choice C would have won in (i) by 4 for C and 3 for A and B.
iv. One shows "disapproval" for those he or she does not vote for in approval voting.

9. i. B beats A by 5 to 4 on the first round, and B then wins over C by 7 to 2 in Agenda I.
B beats C by 7 to 2 on the first round, and B then wins over A by 5 to 4 in Agenda II.
ii. In Agenda I, these four voters cannot improve the outcome for themselves by voting insincerely.
In Agenda II they can vote for their last choice C on the first round. Then C beats B by 6 to 3 (presuming C's two supporters vote sincerely) on the first round, and A wins over C by 7 to 2 on the second round.

SHOW TWO: MORE EQUAL THAN OTHERS: WEIGHTED VOTING

In most voting situations, each voter possesses an equally weighted vote: the case of one man-one vote. Nevertheless, there are many elections in which some voters cast more heavily weighted ballots than others. The purpose of such weighted voting is to undo some inequity at another level. A stockholder with more shares should have more influence on a corporation's decisions. A larger state should have a greater say about who should be the President of the U.S.A. A county board of representation may have representatives elected from equal sized districts who each cast one vote on the board. On the other hand, the representatives on a county board may each represent different smaller municipalities with unequal populations. In this latter case, it seems reasonable for the representatives from larger districts to have a more heavily weighted vote than those from smaller constituencies.

The primary lesson in this show is to see that the influence of one's vote need not be proportional to the weight of his or her weighted vote. One's power need not have a simple relationship to the fraction of the voters' weights. One such measure of power is called the *Banzhaf power index*. It counts the number of voting combinations in which an individual's vote is critical to the outcome. The main goal of this show is to understand and to be able to compute the Banzhaf index for some simple weighted voting systems.

Determining equitable weights to achieve a fair weighted voting system is only one topic in a new mathematical subject concerned with fair division of gains or losses. The apportionment problem for rounding fractions in an equitable manner and more general fair division schemes for dividing rather arbitrary objects (e.g., a chocolate cake) are examples of other topics in the theory of fair division. These topics are mentioned only briefly in this show (and are covered in more detail in Chapter 11 in the text).

Skill Objectives

1. Appreciate the need for weighted voting.
2. Understand how weighted voting works.
3. Learn some vocabulary and concepts about voting.
4. Identify some basic ideas fundamental to the notion of power.
5. Know what the Banzhaf power index is and how to compute it.
6. See an application of elementary set theory.
7. Obtain some practice working with combinations (or hypercubes).
8. Have a glimpse into the nature of some other fair division problems: the apportionment problem and a fair division scheme.

Self-test

1. Give the cardinality of the following sets:
   i. {1, 2, 4}.
   ii. The set of all vertices of a cube.
   iii. The set of all subsets of the set {0, 1}.

2. In the three-person weighted voting game $[q: w_1, w_2, w_3] = [5: 6, 3, 1]$ are any of the voters dictators or dummies?

3. For the weighted voting game [5: 3, 2, 1] list all of the:
   i. winning coalitions.
   ii. minimal winning coalitions.
   iii. losing coalitions.

4. In the weighted voting game [5: 3, 2, 1] given in **Problem** 3 above, how many different ways can voter 1 join a losing coalition and thus turn it into a winning coalition?

5. A college committee has four members: the dean of the college, one professor, one student, and a librarian. Three of the four members must approve any motion before it can pass. However, the dean by herself can veto any issue. Express this voting system as a weighted voting game.

ANSWERS
1. (i) 3, (ii) 8, (iii) 4; 2. Voter 1 is a dictator, and voters 2 and 3 are dummies; 3. (i) {1, 2, 3} and {1, 2}, (ii) {1, 2}, (iii) {1, 3}, {2, 3}, {1}, {2}, {3}, and $\emptyset$; 4. Two ways: {2} to {1, 2} and {2, 3} to {1, 2, 3}; 5. [4: 2, 1, 1, 1].

Sample Problems

6. i. List all the possible combinations of votes for three voters 1, 2, and 3 where each voter can vote either yes (Y) or no (N).
   ii. How many of the combinations in (i) have two Y votes and one N vote?

7. Consider the four-person weighted voting game [6: 4, 3, 2, 1].
   i. List all of the winning coalitions.
   ii. List all of the minimal winning coalitions.
   iii. Is voter 4 a dummy?
   iv. Does voter 1 have more power than voter 2?

8. Give the Banzhaf power index for the three-person weighted voting game [5: 3, 2, 1]
   i. by listing the eight combinations of yes (Y) or no (N) votes, and circling the critical voters in each combination.
   ii. by drawing the lattice (cube) figure of all subsets of {1, 2, 3} and by labelling each

critical edge on this cube by the appropriate pivotal player.

9. A well-known two-person fair division scheme is: one person cuts the "cake" into two pieces, and the other person chooses one of these pieces. Discuss precisely what it is that is being assumed when one asserts that this is a fair division scheme, i.e., what conditions are necessary in order that both the cutter and the chooser will be guaranteed to obtain an acceptable piece?

ANSWERS

6. (i) YYY, YYN, YNY, NYY, YNN, NYN, NNY, NNN; (ii) 3.

7. (i) {1, 2, 3, 4}, {1, 2, 3}, {1, 2, 4}, {1, 3, 4}, {2, 3, 4}, {1, 2} and {1, 3};
(ii) {2, 3, 4}, {1, 2} and {1, 3}.
(iii) No
(iv) Yes

8. (i)

Ⓨ	Ⓨ	Y	Pass
Ⓨ	Ⓨ	Y	P
Y	Ⓝ	Y	Fail
Ⓝ	Y	Y	F
Y	Ⓝ	N	F
Ⓝ	Y	N	F
N	N	Y	F
N	N	N	F

(ii)

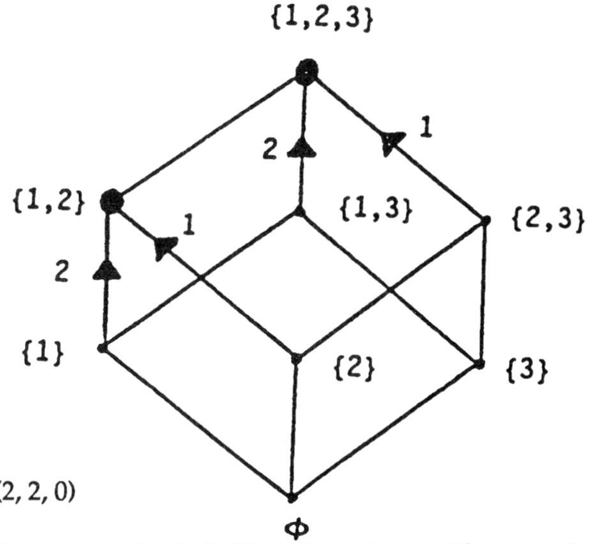

The Banzhaf index is (2, 2, 0)

9. It assumes that the cutter can cut the "cake" into two pieces, *either* one of which is acceptable to her; and given *any* division of the "cake" into two pieces, at least *one* of them will be acceptable to the chooser.

SHOW THREE: ZERO-SUM GAMES: GAMES OF CONFLICT

The mathematical subject called *game theory* was created to study situations involving two or more individuals (or groups) in conflict or cooperation. This show is concerned with only the two-person, zero-sum games. These are presented by a table of numbers and are thus referred to as matrix games. The term "zero-sum" means the players are in total conflict; a gain for one is an equivalent loss for the other. Their objectives are in complete opposition. Many such games occur in warfare, sports, and parlor games. The protagonists may be duelists, bidders, inspector and diverter, searcher and evader, or potential terrorist and security agent.

A matrix game consists of two players, a list of choices for each player (called strategies), and a tabulation of the outcomes (called payoffs) resulting from the selection of particular strategies. One player is called the row player and his strategies correspond to the rows of the matrix, and the other player is the column player whose strategies are the columns of the matrix. The numbers in the matrix are the resulting payoffs won by the row player and lost by the column player.

For some games, like the restaurant location game, it is rather straight forward to determine the optimal strategy for each player and the corresponding outcome of the game. Such games are said to have a *saddle point*. Other games such as the illegal parker versus the policeman do not have a saddle point in the original (pure) strategies. To solve the latter type of games, one must introduce the concept of a mixed strategy. This is a probability distribution over one's pure strategies. The optimal solution is then given in terms of the best such randomized strategies, and the value of the game is then taken as the *expected value* in a statistical sense.

To appreciate the notion of a mixed strategy and the solution of these games, one should understand the notion of probability (or mean in statistics). To actually solve to find an optimal mixed strategy for any game when one of the players has only two pure strategies, one must be able to solve two linear equations in two unknowns. This prerequisite from high school algebra is similar to the background needed to solve the simple linear programming problems appearing in an alternate show.

Skill Objectives

1. Recognize some competitive game situations.
2. Represent a two-person, zero-sum game by a matrix.
3. Learn some basic concepts and vocabulary from game theory.
4. To find the optimal pure strategies and value of a game with a saddle point by means of the minimax technique.
5. Appreciate the concept of and the need for a mixed strategy.
6. To compute an optimal mixed strategy and the (expected) value for a player in a game who has only two pure strategies.

Self-test

1. Which of the following matrix games has a saddle point?

i. $\begin{bmatrix} 1 & 5 \\ 4 & 2 \end{bmatrix}$ ii. $\begin{bmatrix} 2 & 2 & 5 \\ 3 & 1 & 4 \end{bmatrix}$

iii. $\begin{bmatrix} -1 & 5 \\ 3 & 4 \\ 1 & -2 \end{bmatrix}$ iv. $\begin{bmatrix} -2 & 2 & -3 \\ -1 & 0 & -1 \\ -1 & 4 & -2 \end{bmatrix}$

2. Give the value of the games in **Problem 1** which do not have a saddle point.

3. Give the optimal (pure) strategies for the games in **Problem 1** which do have a saddle point.

4. For any game in **Problem 1** which does not have a saddle point, give the maximin value for the row player and the minimax value for the column player (i.e., the best payoffs these players can be assured of obtaining using pure strategies).

5. Solve the restaurant location problem is the game matrix is given as follows.

		Lisa		
		1	2	3
Henry	*A*	8	7	6
	B	9	5	5
	C	4	8	4

ANSWERS
1. (ii), (iii), and (iv).
2. (ii) 2, (iii) 3, and (iv) -1.
3. (ii) row player: 1, column player: 2;
   (iii) row player: 2, column player: 1;
   (iv) row player: 2, column player: 1 and 3.
4. (i) maximin is 2 for row 2 and minimax is 4 for column 1.
5. There is a saddle point at row A and column 3 with value 6.

Sample Problems

6. Two bidders can bid in multiples of one dollar on an object worth \$6. The higher bidder wins the object and pays the amount of his bid to the lower bidder. If there is a tie, then they flip a coin to determine the winner, which they each evaluate as worth \$3 (half of the \$6). However, the row player only has four dollars in hand to bid with. This two-person (constant-sum) game can be represented by the following matrix.

(The column player's winnings are $6 minus the payoffs in this matrix.)

		Column Player's Bids						
	$	*0*	*1*	*2*	*3*	*4*	*5*	*6*
	0	3	1	2	3	4	5	6
Row	1	5	3	2	3	4	5	6
Player's	2	4	4	3	3	4	5	6
Bids	3	3	3	3	3	4	5	6
	4	2	2	2	2	3	5	6

Find the value and any optimal (pure) strategies for this game. Is the row player at a disadvantage because he or she has fewer dollars to bid with?

7. Describe the payoff matrix for the following two-person, zero-sum coin matching game. Each player can show one coin: a penny, a nickel or a dime. If the sum of the values of the two coins shown is an odd number then the row player wins the column player's coin. If the sum is even then the row player loses his coin to the column player.

8. Solve the following matrix game.

$$\begin{bmatrix} -3 & 5 & -2 & 2 \\ -2 & 1 & 0 & -1 \\ -4 & -3 & 3 & 4 \end{bmatrix}$$

9. Solve the following matrix game for its value and the optimal (mixed) strategies for each player.

$$\begin{bmatrix} 3 & -2 \\ -1 & 3 \end{bmatrix}$$

10. Each one of the two political candidates running in an election must take a stance as either left (L), center (C), or right (R). They estimate the electorate as being 30% L, 30% C and 40% R; and assume that the citizens will vote for the candidate whose stance is closest to their position (or split their vote evenly in the case of a tie).
i. Formulate this problem as a matrix game with the payoffs to the row player being the percentage of the votes he or she wins minus 50%.
ii. Solve this matrix game in (i).
iii. Assume that the candidates are not free to select any stance. The row player must select L or C, and the column player can only pick R. Solve this restricted game.

11. Solve the matrix game in **Problem 2**. Note that this three by three matrix game can

be reduced to a two by two matrix by eliminating one poor strategy for each player.

ANSWERS

6. Value = 3; optimal pure strategies are to bid either 2 or 3 dollars. There is no advantage to having more than 3 dollars to bid.

7.

	penny	nickel	dime
penny	-1	-1	10
nickel	-5	-5	10
dime	1	5	-10

8. Value = -2; optimal pure strategies are row 2 and column 1.

9. The optimal mixed strategy $(1 - q, q) = (5/9, 4/9)$ for the column player is gotten by solving $3(1 - q) - 2q = -1(1 - q) + 3q$ (= value = 7/9). Optimal for the row player is $(1 - p, p) = (4/9, 5/9)$.

10. (i)

	L	C	R
L	0%	-20%	-5%
C	20%	0%	10%
R	5%	-10%	0%

(ii) Value = 0%; optimal strategy is C.
(iii) Value = 10%; row player selects C.

11. Nickel is a poor strategy. The optimal mixed strategy for each player is $(1 - p, 0, p) = (10/11, 0, 1/11)$, and the (expected) value = 0.

SHOW FOUR: PRISONER'S DILEMMA: GAMES OF PARTIAL CONFLICT

In the matrix games of complete conflict discussed in Show Three, the two participants had goals that were directly opposed to each other. In many other game situations, the aims of the players involve a delicate mix of competition and cooperation. Some illustrations of such games, referred to as *games of partial conflict*, are presented in Show Four. These sorts of encounters arise frequently in economic, political, and social interactions. The analysis of such games provides helpful insight into administration, regulation, business, decisionmaking, and even religion. Two rather troublesome games of partial conflict are described in some detail. These games clearly illustrate two common "social paradoxes" which arise routinely in everyday behavior decisions.

The payoffs to the players in a game of partial conflict will not sum to zero. To describe such a two-person game, it is necessary to give two payoffs for each outcome, one for each of the players. So, two numbers must appear in each position of the game matrix. (Alternately, one could use two matrices, one for each player, with single entries for payoffs. Thus these games are often referred to as *bimatrix games.*) Both examples presented here are limited to the case where each player has only two strategies. One choice corresponds to social cooperation, whereas the other option represents a defection from an explicit agreement or social norm which would have been best for the pair when considered together.

In the famous Prisoner's Dilemma game, each of two partners in crime has the choice of either steadfastly maintaining their innocence or else turning state's witness and squealing (i.e., blaming the crime) on his or her partner. This same game is at the heart of many types of escalations such as price wars, arms races, rising trade barriers over exploitation, etc. The other game is Chicken, where at least one of the two proud participants must back down from the brink in order to avoid mutual disaster.

Fortunately, most games of partial conflict (as illustrated in some of the subsequent examination questions) are less bothersome than the two examples mentioned. Even in the games of Chicken and the Prisoner's Dilemma, the socially worst outcome can typically be avoided if participants demonstrate mutual trust. Furthermore, if these games are played repeatedly over time, the players soon learn that it is to their long range interest to cooperate at each play. This idea of the evolution of cooperation over time is discussed and illustrated in classroom experiments by Robert Axelrod.

The only new technical solution concept related to this show is that of an equilibrium outcome. A pair of particular strategies, one for each player, leads to an *equilibrium* when neither *one* of the players can unilaterally change only his or her strategy and thus obtain a better payoff.

It should be noted that video tapes for the *two* shows 3 and 4 correspond to only *one* chapter (Chapter 12) in the textbook.

Skill Objectives

1. Model some simple two-person games of partial conflict (the two-by-two, bimatrix games).
2. Understand the nature of and the implications of playing the game called the Prisoner's Dilemma.
3. Understand the game called Chicken.
4. Appreciate the prevalence of these two games in routine social activities.
5. Recognize when a pair of strategies for two players is in equilibrium.

Self-test

1. Describe any pair of (pure) strategies, one for each player, which is an equilibrium pair in the following two-person games of partial conflict. (Assume that *both* players prefer to receive payoffs which are higher numbers over those that are lower.)

i. $\begin{bmatrix} (4, 1) & (2, 2) \\ (3, 3) & (1, 4) \end{bmatrix}$ ii. $\begin{bmatrix} (4, 1) & (1, 2) \\ (3, 4) & (2, 3) \end{bmatrix}$

iii. $\begin{bmatrix} (2, 2) & (2, 2) \\ (1, 1) & (2, 2) \end{bmatrix}$ iv. $\begin{bmatrix} (4, 1) & (2, 3) \\ (1, 4) & (3, 2) \end{bmatrix}$

2. If either player made use of the maximin technique (from Show Three), then he or she would arrive at a maximum value (which is the worst payoff this player could receive if he or she played accordingly). Find the maximin values and corresponding strategies for both the row and column players in the games given in **Problem 1.**

3. Discuss briefly how the following two situations existing between two countries or two manufacturers can be modeled as a Prisoner's Dilemma game.
   i. trade barriers
   ii. a price war

4. Model a labor negotiation as a game of Chicken where labor's options are to strike if treated unfairly or not to strike, and management's choices are to hold to a hard line or else to offer a fair settlement.

ANSWERS

1. (i) (2, 2) at row 1 and column 2; (ii) none; (iii) (2, 2) at row 1 and column 1, at row 1 and column 2, and at row 2 and column 2; (iv) none.

2. (i) 2 at row 1, 2 at column 2; (ii) 2 at row 2, 2 at column 2; (iii) 2 at row 1, 2 at column 2; (iv) 2 at row 2, 2 at column 2.

3. (i) The cooperative strategy is to maintain free trade, and the defection strategy is to increase barriers (limits or duties) on imports. (ii) The cooperative strategy is to set

"reasonable" prices, and the defection strategy is to attempt to underprice your competitor.

4. One model:

		Management: Hard line	Management: Fair offer
Labor	Strike if must	(1, 1)	(4, 2)
	Do not strike	(2, 4)	(3, 3)

Sample Problems

5. Discuss what strategies you believe rational players would play, and why, in the following games of partial conflict. Assume that the players are playing in a noncooperative mode where they know the payoffs but must choose their strategies in ignorance of their opponent's choice and without prior communication.

i. $\begin{bmatrix} (4, 4) & (2, 2) \\ (3, 3) & (1, 1) \end{bmatrix}$

ii. $\begin{bmatrix} (4, 2) & (2, 3) \\ (3, 4) & (1, 1) \end{bmatrix}$

iii. $\begin{bmatrix} (2, 2) & (1, 2) \\ (2, 1) & (1, 1) \end{bmatrix}$

iv. $\begin{bmatrix} (4, 1) & (2, 3) \\ (1, 4) & (3, 2) \end{bmatrix}$

6. If the two players were able to communicate and make binding agreements before playing the following game called "battle of the sexes," what would you suggest they do?

Plan to go to:		She: Boxing	She: Ballet
He	Boxing	(4, 2)	(1, 1)
	Ballet	(3, 3)	(2, 4)

7. It is recommended that we inoculate infants against diseases such as pertussis (whooping cough), even though only a few infants react negatively to the serum. For example, in 1984, only eight children in the U.S.A. died of pertussis, whereas 38 died from reactions to the vaccination. If given the choice, what should a parent do? Can you describe this vaccination game as a two-person game of partial conflict with one child versus society as a whole being the two players?

8. Consider the three-person coin matching game. Each of the three players 1, 2, and 3 shows heads (H) or tails (t) and the "odd man" wins 25 cents from each of the other two players. If three heads or tails are displayed, it is a tie and no one wins anything. Assume that players 2 and 3 cheat by forming a coalition and planning a joint

strategy to use against player 1.
i. Model this as a two-person game (of conflict) where player 1 plays against the "player" consisting of the coalition made up of 2 and 3.
ii. What strategies will these players use in the two-person game in (i)?

ANSWERS

5. (i) row 1 and column 1; (ii) row 1 and column 2, since row 1 dominates row 2 and the column player's best response is column 2; (iii) row 1 and column 2, for each player's payoff is in the "hands of the other" and you hope for good will on the part of the other player; (iv) row 1 and column 2 for this is a saddle point for this game of conflict.

6. In a single play of the game, they may agree to flip a coin to determine whether to go together to boxing or ballet. If they date frequently, they may agree to alternate which event they attend.

7.

		Society	
		Vaccinate	Do not Vaccinate
Child	Vaccinate	(3, 4)	(2, 2)
	Not	(4, 3)	(1, 1)

Equilibrium at society vaccinate, but individual child not!

8. (i)

		{2, 3}		
		HH	HT	TT
1	H	(0, 0)	(-25, 25)	(50, -50)
	T	(50, -50)	(-25, 25)	(0, 0)

(ii) The coalition {2, 3} will always play one H and one T (column 2) and collect 25 cents from player 1 on each play independent of what player 1 does.

Part IV On Size and Shape
Paul Campbell, Beloit College
Donald Albers, Menlo College
Donald Crowe, University of Wisconsin
Seymour Schuster, Carleton College
Maynard Thompson, Indiana University

INTRODUCTION

Mathematics is the study of patterns and relationships. Mathematicians instinctively search for geometrical and numerical patterns and for symmetry. Their discoveries of patterns and symmetries often enable us to better understand practical problems. Geometry grew as a body of knowledge because of people's need to explain and control the world around them. The set of tools created has enabled mankind to range further and further from daily experience.

From determining the circumference of the earth to digging straight tunnels, the ancient Greeks used the tools of geometry to make seemingly impossible measurements. More modern astronomers and mathematicians succeeded in uncovering the secrets of planetary motion. And new geometries, which have led to a much deeper understanding of the nature of mathematics itself, paved the way for the discoveries of Einstein, the basis of our present view of the workings of our universe.

Perhaps no other subject has intrigued people through the centuries as much as geometry. We perceive the symmetries and patterns of nature and search to understand the intrinsic order and beauty we observe.

OVERVIEW SHOW

Mathematicians are particularly attracted to patterns and relationships. The pattern might be geometrical or numerical. Often geometrical patterns are strongly linked to numerical patterns. As the chambered nautilus grows, each of its new chambers is geometrically similar to the previous chambers, and the chambered curve generated has the form of a logarithmic spiral. The beautiful spiral growth patterns of sunflowers also turn out to be examples of a famous numerical pattern, namely, the Fibonacci sequence, which begins 1, 1, 2, 3, 5, 8, 13, 21, 24, ... Notice that each term is the sum of the previous two terms.

If we look at ratios of successive terms in the Fibonacci sequence, after 3, we see that 5:3, 8:5, 13:8, 21:13, 34:21, ... all are about the same, namely, 1:1.6. Euclid called this ratio the Golden Ratio.* It turns out to be the ratio of the sides of the rectangle that is most pleasing to the eye. That rectangle is called the Golden Rectangle. It can be found in the design of the Parthenon and the Gardens of Versailles. The Golden Rectangle is a manifestation of our search for harmony or proportions.

* The exact value of the Golden Ratio is $1:(1 + \sqrt{5})/2$.

Symmetry is a basic aspect of our attempts to comprehend order and beauty. Mathematicians describe a variety of kinds of symmetry by using the geometric tool of rigid motion, or *isometry,* which means "of the same measure." A rigid motion is a transformation in the plane (or in space) in which the original figure and its image are congruent. In this show, symmetry is examined as it arises from four fundamental motions: translation, rotation, reflection, and glide-reflection. These transformations have proven to be very important in classifying the symmetries found in repeating patterns of pottery and artwork of lost cultures.

The Overview concludes with a look at the interactions and applications of two relatively new geometrical ideas--non-Euclidean geometry and fractals. Non-Euclidean geometries arise when the Parallel Postulate is replaced with one of two alternative postulates:

1. Given a line and a point not on the line, infinitely many parallels to the line may be drawn through the point.
2. Given a line and a point not on the line, no parallels to the line may be drawn through the point.

Non-Euclidean geometry turned out to be very important to the Dutch artist Escher, enabling him to draw intricate patterns of 2- and 3-dimensional objects. However, a large class of objects is not so easy to describe. Natural objects such as mountains, clouds, subatomic structures, and planets have forms that are either amorphous or very complex.

The concept of fractional dimension, or *fractal,* was developed in order to describe the shapes of natural objects. In geometry, we consider a line a one-dimensional object and a plane a two-dimensional object. What dimension should a jagged line have? Within the concept of fractals, a jagged line is given a dimension between 1 and 2, with the exact value determined by the line's "jaggedness." Fractals have a host of important applications including the production of spectacular computer images.

SHOW ONE: HOW BIG IS TOO BIG: SCALE AND FORM

This show explores two basic ideas: *geometric similarity* and *tiling*. Both ideas have surprisingly strong ties to the real world. You recall that two objects are similar if they have the same shape. Notice that similar objects do not need to have the same size. Similarity, when combined with a few biological and physical facts, is essential to understanding why things are the size that they are. It explains why the giants and tiny people of *Gulliver's Travels* are only fictional creatures--not to be found in real life. It also forbids the existence of monsters like King Kong and Godzilla. Similarity also is fundamental to understanding why the tallest trees on earth cannot be much higher than 330 feet, and why the tallest mountains can only be seven miles high.

The key to understanding most of these somewhat surprising results is:

1. the *surface area* of a scaled-up object increases as the *square* of the scaling factor and
2. the *volume* of a scaled-up object increases as the *cube* of the scaling factor.

As an example, consider an ordinary mouse about 2" long and a fictional monster mouse, *similar* to the ordinary mouse and 100 times its size. The surface area of the monster mouse would be 100 x 100 = 10,000 times that of the ordinary mouse, but the volume of the monster mouse would be 100 x 100 x 100 = 1,000,000 times that of the ordinary mouse. This would mean that the weight of the scaled-up mouse was one million times that of the ordinary mouse. Since bone strength is determined by cross-sectional area, the bones of the monster mouse would crumble under its own weight.

The second big idea of this show is *tiling*. The use of repeated shapes to cover a plane surface without gaps or overlaps is called a *tiling*. An especially common tiling found on kitchen floors and bathrooms is that done with squares, all the same size.

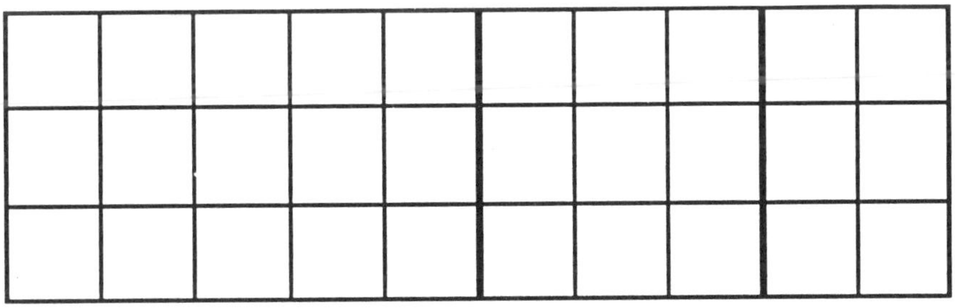

Equilateral triangles also *tile* the plane as do *regular* hexagons. (Regular means all angles and sides are equal in size.) If we restrict ourselves to regular polygons and edge-to-edge patterns, then only three regular polygons can tile the plane--the triangle, the square, and the hexagon.

Very often, tilings are made up of two or more types of regular polygons. Take, for example, the tiling in **Figure 1** that is composed of hexagons and squares. To the surprise of rug designers only twenty-one, semi-regular tilings are possible.

Figure 1.

Show 1 concludes by examining tilings that do not repeat and were believed not to be found naturally in crystals. Within the past few years, some non-periodic tilings have been found in the laboratory, which violated long held "laws" of crystallography.

Skill Objectives

1. Learn the idea of geometric *similarity*.
2. Learn how surface area and volume depend upon the *scaling factor*.
3. Learn the idea of *tiling*.
4. Learn the idea of *regular polygon*.
5. Be able to recognize *regular, semi-regular,* and *periodic* tilings.

Self-test

MULTIPLE CHOICE

1. Which of the following pairs of objects are geometrically similar?
   a. square, triangle
   b. triangle, pyramid
   c. red square, red triangle
   d. ping pong ball, beach ball

2. If a sphere is scaled up by a factor of 4, then its volume is
   a. 64 times the volume of the original sphere.
   b. 8 times the volume of the original sphere.
   c. 16 times the volume of the original sphere.
   d. 4 times the volume of the original sphere.

3. If a sphere is scaled up by a factor of 10, then its surface area is
   a. 10 times the surface area of the original.
   b. 20 times the surface area of the original.
   c. 50 times the surface area of the original.
   d. 100 times the surface area of the original.

4. Limitations on the heights of mountains on the earth are due primarily to
   a. sunspots.
   b. regular tilings.
   c. trans-molecular scaling.
   d. pressure/tensile strength considerations.
   e. all of the above.

5. Skyscrapers are *not* built out of wood because
   a. of excessive dry rot problems.
   b. of cost considerations.
   c. not enough wood is available.
   d. wood is not strong enough to support the weight of skyscrapers.

6. The sail on the Dimetrodon's back
   a. was essential to increasing its speed on water.
   b. was decorative and served no useful purpose.
   c. evolved as a way to give him more surface area.
   d. changed over time to provide him with more heat as the earth cooled.

7. Regular tilings of the plane can be done with all of the following except
   a. triangles.
   b. squares.
   c. pentagons.
   d. hexagons.

8. The pattern below is an example of

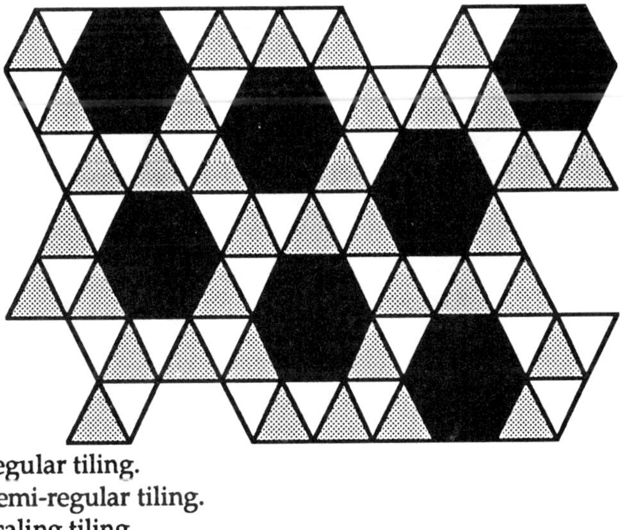

   a. a regular tiling.
   b. a semi-regular tiling.
   c. a scaling tiling.
   d. a non-periodic tiling.

9. Non-periodic tilings
   a. are forbidden by laws of similarity.
   b. were first found in the laboratory.
   c. are the same as semi-regular tiling.
   d. were first discovered by mathematicians.

10. Penrose tiling exhibits
   a. fivefold symmetry.
   b. vertex ambivalence.
   c. rotational instability.
   d. dart-kite skewness.

ANSWERS
1. (d), 2. (a), 3. (d), 4. (d), 5. (d), 6. (c), 7. (c), 8. (b), 9. (d), 10. (a).

Sample Problems

11. An object has a surface area of 4 square inches and a volume of 2 cubic inches. If it is scaled up by a factor of 5, find the new surface area and volume.

12. How high could a mountain be on Mars, whose gravity is about 0.38 times that of Earth? (The theoretical figure for the tallest mountain on Earth is 7 miles.)

13. a. "Show" that the triangle below tiles the plane.
    b. Do the same for the quadrilateral shown below.

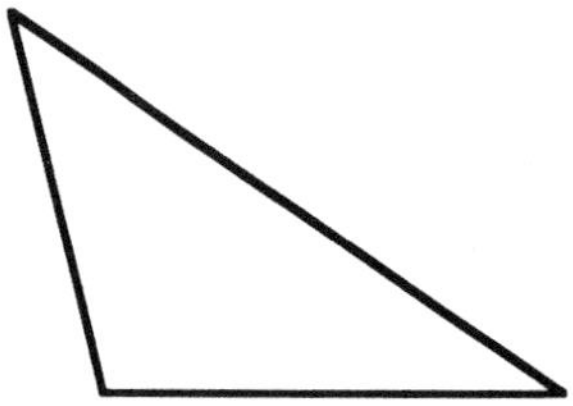

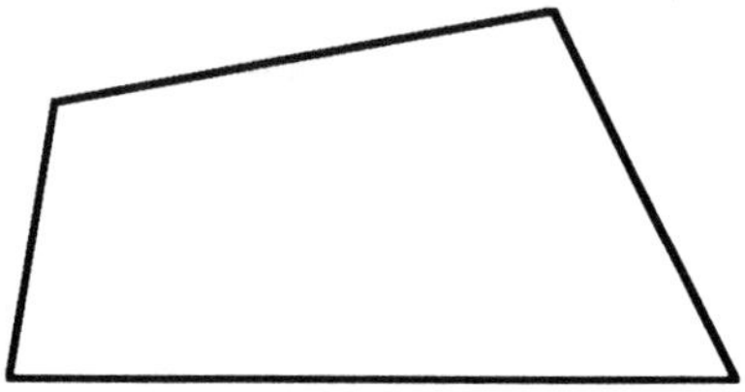

14. Determine the size of each interior angle of a regular 12-gon.

ANSWERS
11. Surface area increases with the square of the scaling factor. Since the scaling factor is 5, it's square is 25. Thus the surface area of the scaled-up object is 25 x 4 sq. in. = 100 sq. in.
    Volume increases with the cube of the scaling factor. Thus the volume of the scaled-up object is 5^3 x 2 cubic in. = 250 cubic inches.

12. Since the weight of a object is determined by the force of gravity, that means that objects on Mars are 0.38 times their weight on earth. Since the weight of our theoretical cubical mountain is determined by height, we have

$$\frac{H\ (\text{Mars})}{7\ \text{miles}} = \frac{1}{0.38}$$

Thus $$H\ (\text{Mars}) = \frac{7\ \text{miles}}{0.38} = 18\ \text{miles}$$

(The height of Olympus Mons, the tallest mountain on Mars, is about 15 miles. If we use the "actual" height of Mt. Everest, the tallest mountain on Earth, we would get 6 mi./0.38 = 15.8 miles.)

13.

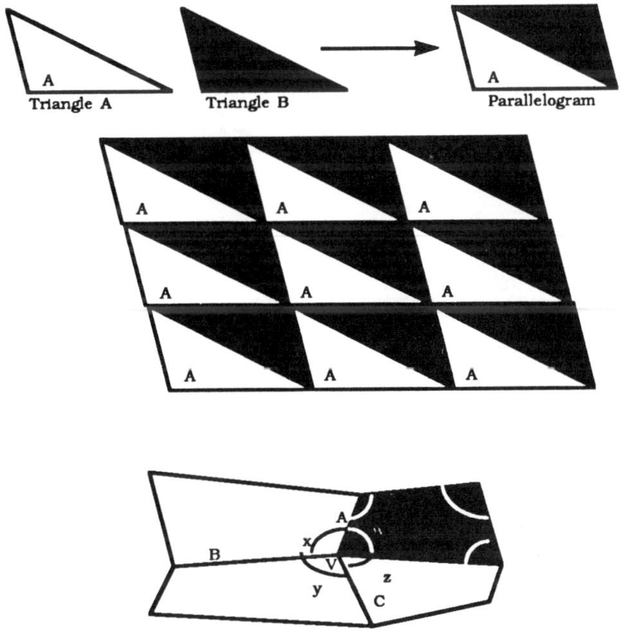

14. A regular 12-gon has 12 equal exterior angles with sum 360°/12 = 30°. Hence each interior angle measures 180° - 30° = 150°.

SHOW TWO: IT GROWS AND GROWS: POPULATIONS

Although it isn't obvious at first blush, money, bunnies, and fish grow in very similar ways. If you deposit \$100 in a bank that pays 10% interest compounded annually, then at the end of one year you have your original deposit of \$100 plus interest of 10% x \$100, which is \$10. Thus your total investment grows to \$110 at the end of one year. At the end of two years you would have \$110 + interest of 10% x \$110, \$11, for a total of \$121. In the second year you got more interest because you earned "interest on your interest," which is the core idea behind compound interest. After n years you \$100 would grow to \$100 x $(1 + .10)^n$.

In a similar fashion, if the current size of a bunny population is 100, and it is growing at an annual rate of 10%, then at the end of n years it would grow to $100 \times (1 + .10)^n$ bunnies, the same total (apart from units) that we got for money. In the first example we earned "interest on our interest" and in the second "new bunnies help to produce more new bunnies."

More generally, if we begin with an initial population P that is growing *geometrically*, like money in a bank subject to compounding, at a rate r per period, then after n periods the population grows to $A = P(1 + r)^n$.

Some banks claim that they compound interest *continuously* ("all of the time"). It can be shown that if you deposit \$100 at 100% interest, compounded continuously, then your \$100 will grow to \$100 x e, \$271.83, after one year. In general, an initial deposit P at interest rate r grows to $P \times e^r$ after one year if interest is compounded continuously. (The number $e = 2.71828\ldots$ is a special number that often turns up in real-world problems.)

For biological populations, growth is constrained by limitations of food supplies and living space. These constraints result in modifications to our basic formula $A = P(1 + r)^n$, and lead naturally to problems of *maximum sustainable yield*. If we know that a population, say fish, is limited, then we need to determine a harvesting policy that doesn't exhaust the population of fish. If in one season we don't harvest more than the population grows, then the fish population shouldn't be depleted. If a fish population grows from 100 tons to 150 tons in one season, then it would be safe to harvest 50 tons and sustain our yield. It might be the case that the same population if initially at 120 tons would grow to 180 tons for a sustainable yield of 60 tons. The trick is to find the population size that gives the *maximum sustainable yield.*

This show concludes by examining *demography*, the study of the growth of human populations, paying special attention to *population pyramids*--graphical representations of population distributions grouped according to age and sex. Population pyramids are essential to planning systems for health care, housing, and schools.

Skill Objectives

1. Learn the idea of *geometric growth.*
2. Know and be able to use the formula for *geometric growth.*
3. Learn the significance of the number e in growth problems.
4. Learn the idea of *maximum sustainable yield.*
5. Learn the idea of *reproduction curve.*
6. Learn the idea of *population pyramid.*

Self-test

MULTIPLE CHOICE

1. The formula for *Geometric Growth* is
   a. $A = Prt$.
   b. $P = Art$.
   c. $A = Pr^n t$.
   d. $A = P(1 + r)^n$.

2. *Geometric growth* describes the growth of
   a. geometrical objects such as triangles.
   b. money in a savings account.
   c. children before puberty.
   d. biological populations only.

3. If you invest \$1000 at 10% interest compounded annually, then after three years you will have
   a. \$1300.
   b. \$3100.
   c. \$100 x e.
   d. \$1331.

4. If you invest \$1000 at 12% annual interest, compounded continuously, then after one year you will have
   a. \$1000 x e.
   b. (\$1000 x e) - 12.
   c. \$1000 x $e^{0.12}$.
   d. \$1000 x (1 + e).

5. The formula for geometric growth of biological populations is usually constrained by limitations of
   a. the money supply.
   b. inflation.
   c. food supplies.
   d. titanium supplies.

6. A harvesting practice in which the amount harvested is balanced by the amount replenished is called a
   a. balanced harvest.
   b. reproductive harvest.
   c. sustained yield policy.
   d. balanced replenishment.

7. The *maximum sustained yield* can be determined if you know the
   a. growth rate.
   b. reproduction curve.
   c. annual yield.
   d. sustained yield.

8. Assuming that the population of the United States is growing geometrically and is 230 million in 1986 with a growth rate of 2%, what will it be in the year 2000?
   a. $230 \times (1 + .02)^{14}$ million
   b. $230 \times e^{.02}$ million
   c. $230 \times (1 + .02) \times 14$ million
   d. $230 \times (.02) \times 14$ million

9. *Population pyramids* are
   a. distributions of populations by age and sex together.
   b. three-dimensional, pyramid-shaped population distributions.
   c. essential to U.S. nuclear policy.
   d. stratified random samples of biological populations.

10. Which population pyramid shown in **Figure 2** shows the highest birth rate?

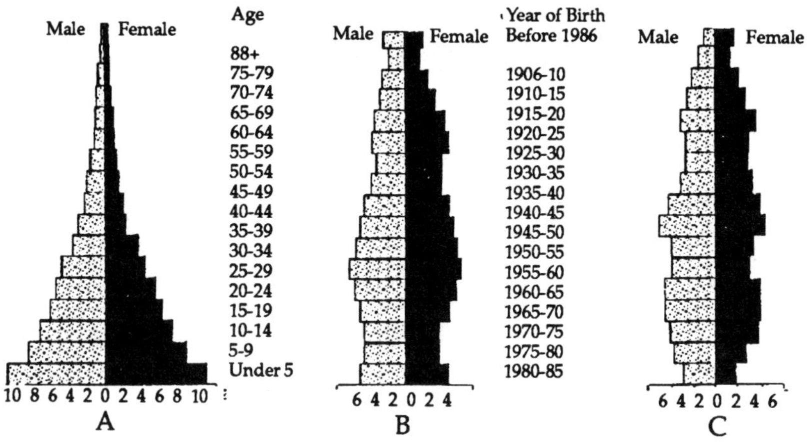

Figure 2.

a. C
b. A
c. B

ANSWERS
1. (d), 2. (b), 3. (d), 4. (c), 5. (c), 6. (c), 7. (b), 8. (a), 9. (a), 10. (b).

Sample Problems

11. Suppose that a country now has a population of 20 million and that its growth rate has been 3% per year for the last ten years. What was the country's population ten years ago?

12. You make an investment of $1000 at 8% annual interest, compounded quarterly. How long does it take for your investment to triple?

13. Suppose a reproduction curve for a certain population is as in **Figure 2.**
  a. Estimate the sustainable yield corresponding to $x = 10$, $x = 30$.
  b. Estimate the maximum sustainable yield.

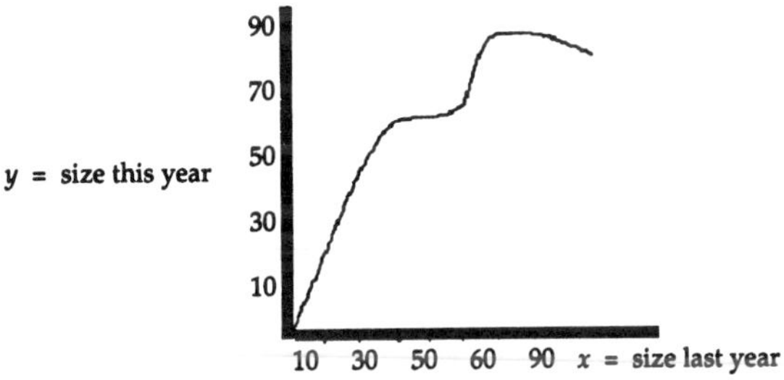

14. Suppose a population of size P is growing at a rate r given by $r = P(100 - P)$. Estimate the value of P corresponding to the largest value of r.

ANSWERS
11. If P was the population size ten years ago, then we have

$P(1 + 0.03)^{10} = 20\text{m}$, thus

$P(1.3439) = 20\text{m}$. Hence,

$$P = \frac{20}{1.3439} = 14.9 \text{ million.}$$

12. After n quarters, your investment will have grown to \$1000 x $(1 + .02)^n$. We seek n for which $\$1000(1 + .02)^n = \$3000$, or $(1 + .02)^n = 3$. Using the x^y key on your calculator, we see after a few tries that n = 56 quarters = 4 years.

13. The sustainable yield is the difference between y values on the reproduction curve and the 45°-line $y = x$. When $x = 10$, the sustainable yield is about 20. The maximum sustainable yield is the largest difference between y values on the reproduction curve and the 45°-line; it appears to be about 30+, occurring at $x = 20$.

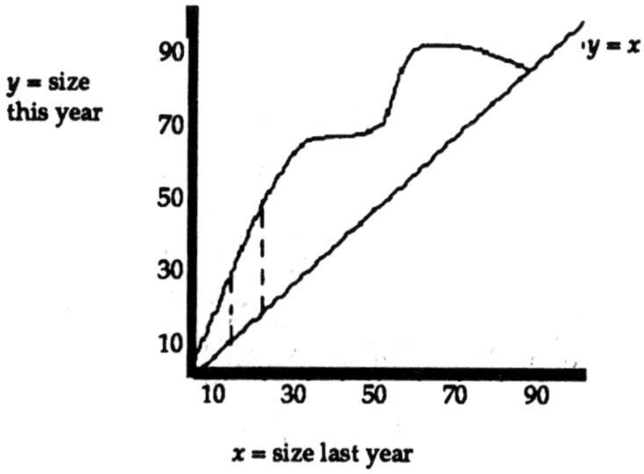

14. If we draw the graph of r versus P, we see that the largest value of r occurs at $P = 50$.

SHOW THREE: STAND UP CONIC: CONIC SECTIONS

Conic sections are very old ideas, but very important. They date back more than 2000 years. About 100 B.C., the Greek mathematician Apollonius wrote eight books about the conics. The conics are the various intersections of a plane slicing a cone. Among those intersections are *circles, ellipses, hyperbolas*, and *parabolas*.* If the slice is in the same direction as the side of the cone, the curve that results is a *parabola*.

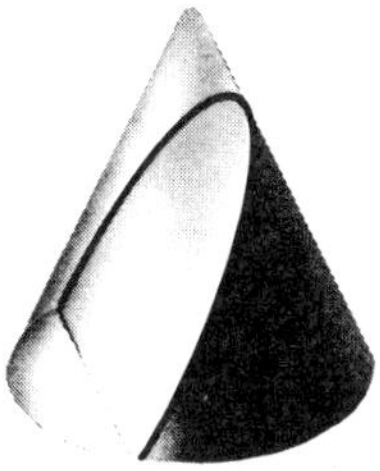

Side view *Perspective view*

Parabola

If the slice is tilted from the direction of the side of the cone toward the horizontal, the curve is an *ellipse* instead.

Side view *Perspective view*

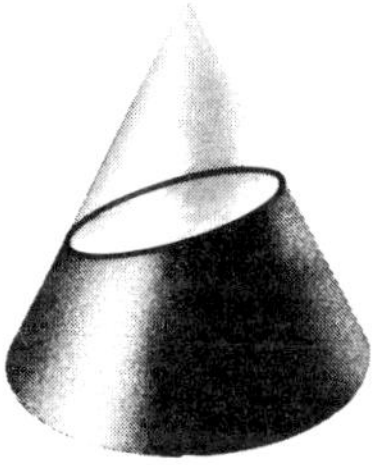

Ellipse

If the slice is tilted in the other direction--that is, away from the direction of the side of the cone and toward the vertical--part of a *hyperbola* is formed.

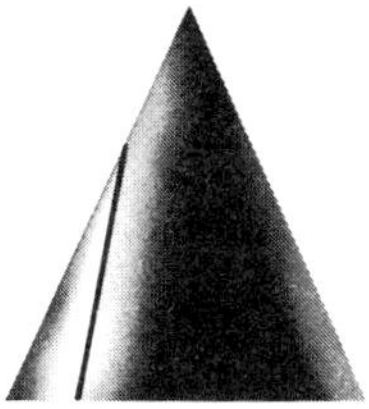

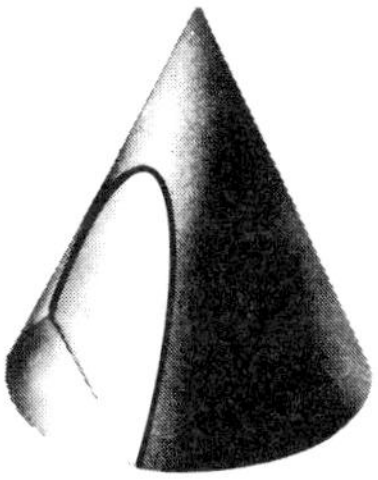

Side view *Perspective view*

Hyperbola

*The full list includes two degenerate cases: the line and a point.

To see the rest of the hyperbola, imagine a second cone balanced upside down on the point of the first one. If the slice is continued through the second cone, the other branch of the hyperbola is formed.

An *ellipse* can also be defined as the set of points in a plane, the *sum* of whose distance from two fixed points is a constant. Each of the fixed points is called a *focus*. It was a major advance in astronomy when Kepler in the 1600's discovered that each of the planets follows an elliptical path with the sun at one focus. Newton went beyond Kepler when he showed that all the celestial bodies influenced only by the sun move in paths that are conic sections.

Kepler also showed that during a given time interval, a line from the sun to the planets sweeps out an equal area anywhere along its elliptical path. (Equal areas are swept out in equal times.) Ellipses have also been used in the design of airplane wings and bicycle gears.

The reflection property of the ellipse is simple: a ray of light which originates from one focus and reflects off the ellipse, will always return to the other focus. This holds true for any kind of ray or for a physical object like a ball. This reflection property explains how elliptically-shaped whispering galleries work. It also explains the reflection properties of elliptically-shaped pool tables.

A *parabola* can also be defined as the set of points in a plane that are equidistant from a fixed point, called the *focus*, and a fixed line, called the *directrix*. The parabola, like the ellipse, is very important to astronomy, but in a very different way. All of the large telescope mirrors in the world are *parabolic*. When parallel light rays from distant celestial objects strike a parabolic mirror parallel to the axis of the parabola, then they reflect through the focus. At the focus, an eyepiece is used to magnify the image brought to the focus. The same idea is used in reverse in automobile headlights. Parabolas also can be seen as the path of projectiles.

A *hyperbola* can also be defined as the set of points in a plane such that the difference between the two distances from each point to two fixed points, called the *foci*, is the same. The *hyperbola*, unlike the parabola and ellipse, scatters sound and light rays. It is thus important in the design of street lamps, flashlights, space heaters, and secondary systems of reflecting telescopes. Hyperbolas are also important in the design of nuclear cooling towers.

Skill Objectives

1. Learn the idea of *conic sections*.
2. Be able to describe how *circle, ellipse, hyperbola,* and *parabola* are defined in terms of cone and cutting plane.
3. Learn Kepler's first law.
4. Learn Kepler's law of "equal areas in equal times" for planets.
5. Learn alternative definitions of ellipse, hyperbola, and parabola.
6. Learn the principle of reflecting telescope.
7. Learn the reflective properties of the conics.

Self-test

MULTIPLE CHOICE

1. All of the following are conic sections except the
   a. ellipse.
   b. parabola.
   c. sphere.
   d. hyperbola.

2. All of the planets have orbits that are
   a. hyperbolas.
   b. parabolas.
   c. ellipses.
   d. circles.

3. Kepler showed that a line from the sun to a planet
   a. covers equal distance in equal times.
   b. varies in length.
   c. sweeps out equal areas in equal times.
   d. is constant in length.

4. If a cone is sliced parallel to one of its sides, the resulting conic section is
   a. an ellipse.
   b. a hyperbola.
   c. a parabola.
   d. a circle.

5. Large reflecting telescopes have primary mirrors that are
   a. hyperbolic.
   b. parabolic.
   c. elliptic.
   d. catadioptric.

6. Whispering galleries make use of
   a. concealed elliptic amplifiers.
   b. the parabolic nature of sound waves.
   c. reflection properties of ellipses.
   d. bifocal magnification.

7. The conic section that results from intersecting both nappes of a cone with a plane is called
   a. an ellipse.
   b. a hyperbola.
   c. a parabola.
   d. an oval.

8. The parabola is important in the design of
   a. automobile headlights.
   b. street lights.
   c. nuclear cooling towers.
   d. bicycle gears.

9. If a light ray emanates from one focus of an elliptical reflector, it
   a. immediately reflects back to the same focus.
   b. reflects to the other focus.
   c. reflects directly between the foci.
   d. reflects through the directrix.

10. The path of every projectile is
   a. an ellipse.
   b. a parabola.
   c. a hyperbola.
   d. a circle.

ANSWERS

1. (c), 2. (c), 3. (c), 4. (c), 5. (b), 6. (c), 7. (b), 8. (a), 9. (b), 10. (b).

Sample Problems

11. Trace the path of a light ray which starts at one focus of the elliptical reflector below. Trace the path through four reflections.

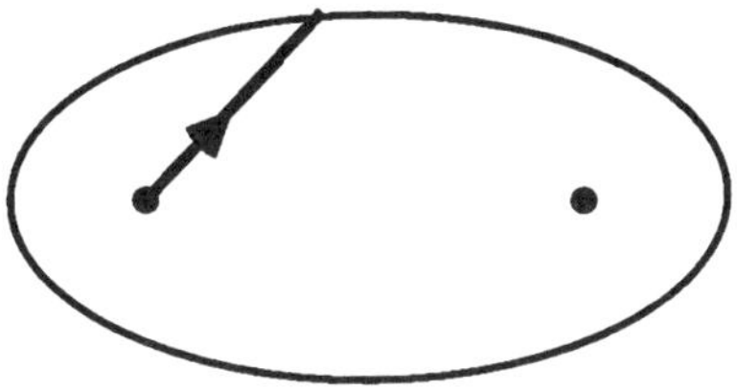

12. Suppose a planet is moving at constant velocity along a straight line path. Show that the line from the sun to the planet sweeps out equal areas in equal times.

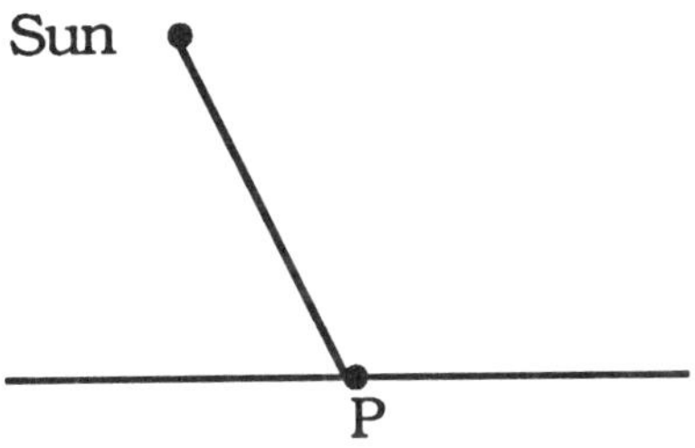

13. Explain how automobile headlights utilize reflection properties of the conics.

14. Ellipses are sometimes referred to as "flattened" ("stretched") circles. The equation $x^2 + y^2 = 1$ represents a circle of radius 1, centered at the origin. Solving for y gives

$$y = + \sqrt{1 - x^2}.$$

Stretching the y's be 2 yields

$$y = + 2\sqrt{1 - x^2}.$$

Show that the latter represents an ellipse.

ANSWERS

11.

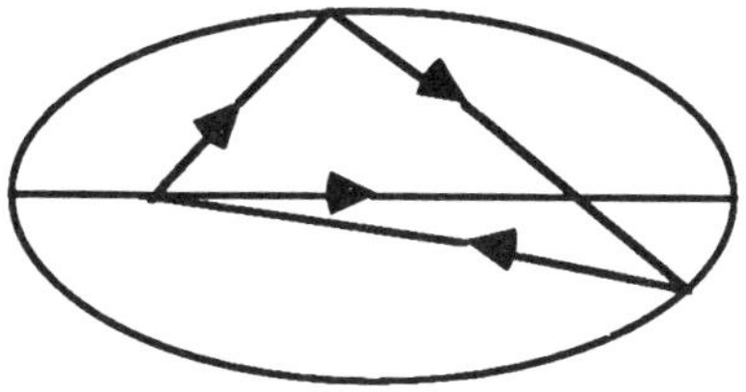

12.

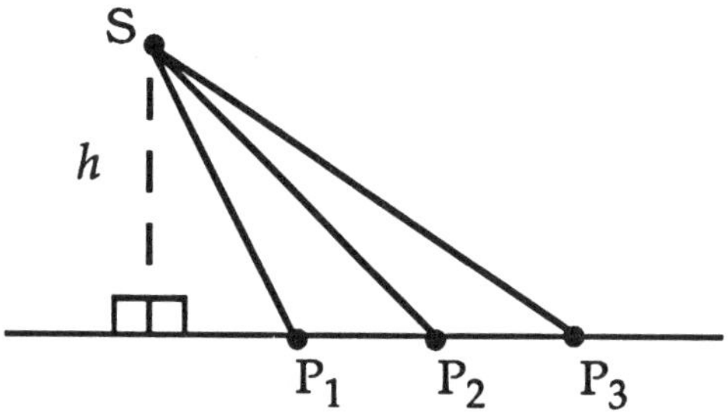

If the planet is moving at constant velocity, then in equal times the distances P_1P_2 and P_2P_3 are the same. Now the triangles ΔSP_1P_2 and ΔSP_2P_3 have the same altitude h. Thus, their areas are equal.

13. Automobile headlights consist of a parabolic reflector and a bulb placed at the focus of the parabola. Light from the bulb strikes the reflector and reflects in a direction parallel to the axis of the parabola. Thus, light from the bulb is concentrated in a narrow beam.

14. If $y = \pm 2\sqrt{1 - x^2}$, then

$$y^2 = 4(1 - x^2)$$

$y^2 = 4 - 4x^2$. Hence,

$4x^2 + y^2 = 4$, and

$x^2 + \frac{y^2}{4} = 1$, which describes an ellipse.

SHOW FOUR: IT STARTED IN GREECE: MEASUREMENT

Remarkable measurements of objects on the earth and of the earth itself can be accomplished using three very old but very powerful ideas of *Euclidean geometry*, *congruence* of triangles, the *Pythagorean* theorem, and *similarity*.

Two triangles are *congruent* if one is an exact copy of the other. That means their corresponding angles and sides are equal. If you go to a copy center and make 10 photocopies of a triangle, then all 10 copies are congruent to each other. In this show we saw how Thales measured the distances of ships from shore by using congruent triangles.

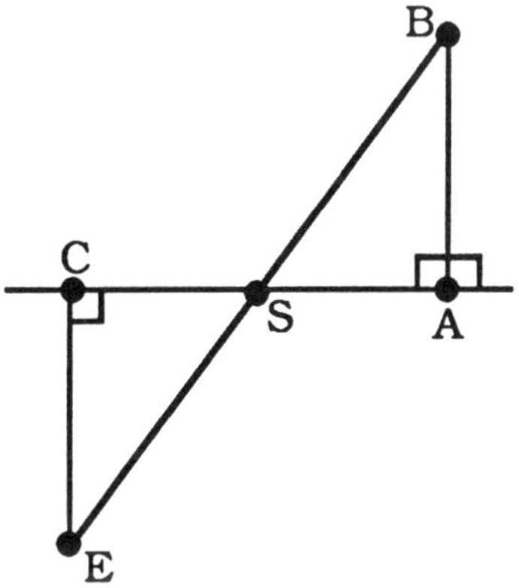

His steps were as follows:

1. Stand directly opposite the Ship B.
2. Pace off the distance *AS* and next pace off a distance *SC* equal to *AS*.
3. Finally turn through 90° at C and walk away from the shore until the boat B lines up with the stake planted at *S*.
4. The distance *CE*, which he paced off, must be equal to *AB*, the boat's distance from shore.

The *Pythagorean* theorem says that in any right triangle the square of the longest side, the hypotenuse, is equal to the sum of the squares of the other two sides.

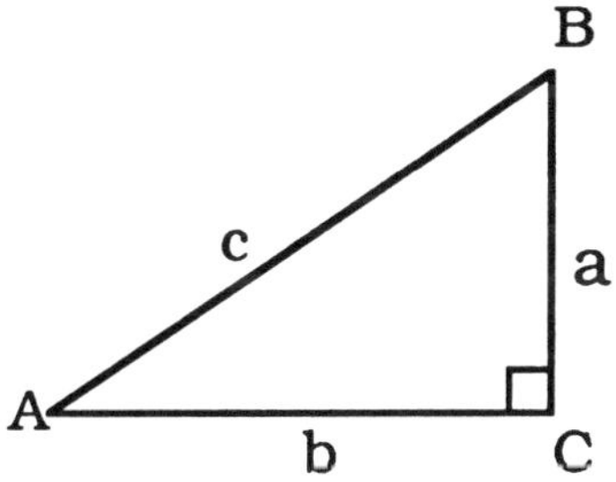

If in a right triangle, the two shorter sides are 5 and 12, then we must have $c^2 = (5)^2 + (12)^2 = 25 + 144 = 169$. Thus $c = \sqrt{169} = 13$.

Two triangles are *similar* if they have the same shape. That means that their corresponding angles are equal and that the ratios of corresponding sides are also the same. Informally, two triangles are similar if one is an enlargement of the other. The two triangles in **Figure 2** are similar because corresponding angles are equal.

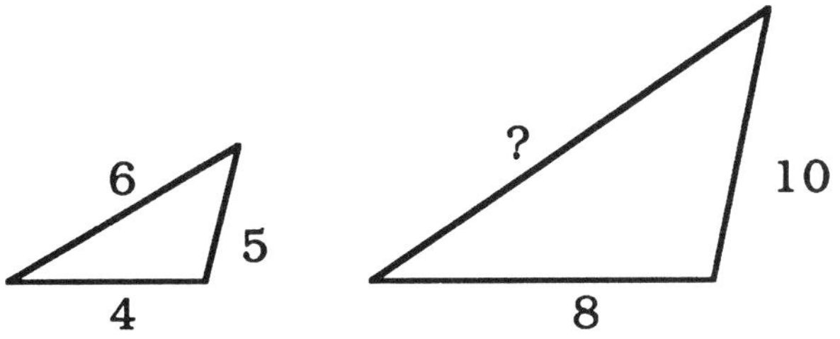

Figure 2.

The unknown side of the second triangle labelled ? must be 12 since the enlargement factor is 2, i.e., the ratio of corresponding sides is 2.

Trigonometry (the measurement of triangles) revolves around the basic idea that the ratios of corresponding sides of similar right triangles are the same. It tells us that in a right triangle, the size of one of the acute angles completely determines the ratios of one side to another. Those ratios have been given special names. For triangle ABC in **Figure 3,**

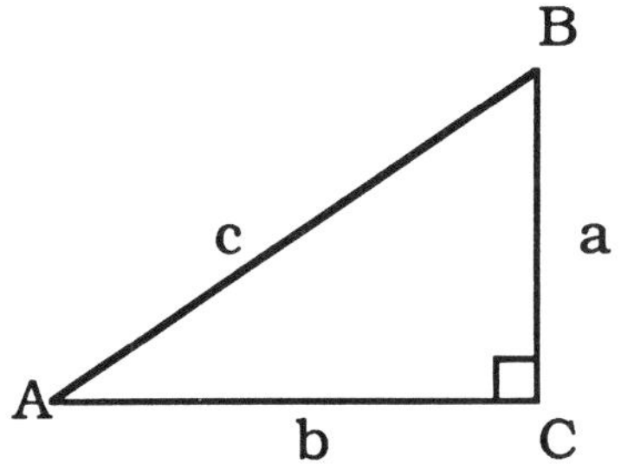

Figure 3.

tangent $A = a/b$, sine $A = a/c$, and cosine $A = b/c$. Tables of these ratios for angles between 0° and 90° are essential to applications.

In the 19th century, Lobachevsky, Bolyai, and Gauss found a new geometry by replacing Euclid's postulate on parallel lines with a new postulate, which allowed more than one line to be drawn through a point P not on a given line. Their work produced a new geometry called *hyperbolic* geometry. In hyperbolic geometry, the sum of the angles of any triangle is less than 180°, and as the area of a triangle increases, so does its angle sum.

By examining a globe we can get some insights into another kind of non-Euclidean geometry, namely *elliptic geometry*. On a globe, we see that there are no parallel lines and that the angle sum of any triangle is greater than 180°. Non-Euclidean geometries are more than an interesting intellectual exercise and have played a vital role in the development of general relativity.

Skill Objectives
1. Learn the idea of *congruence.*
2. Learn the idea of *similarity.*
3. Learn the *Pythagorean theorem.*
4. Be able to apply *congruence, similarity,* and the *Pythagorean theorem.*
5. Learn the *parallel postulate.*

Self-test

MULTIPLE CHOICE

1. If one triangle is an exact copy of another, then they are said to be
   a. Euclidean.
   b. non-Euclidean.
   c. congruent.
   d. similar.

2. Triangles having the same shape are
   a. congruent.
   b. isosceles.
   c. similar.
   d. hyperbolic.

3. The Pythagorean theorem applies to
   a. similar triangles.
   b. congruent triangles.
   c. isosceles triangles.
   d. right triangles.

4. Euclid is best known for
   a. his work on the parallel postulate.
   b. *The Principia.*
   c. hyperbolic geometry.
   d. *The Elements.*

5. The sum of the angles of any spherical triangle is
   a. less than 180°.
   b. equal to 180°.
   c. greater than 180°.
   d. greater than 90° and less than 180°.

6. In spherical geometry, the number of parallel lines that can be drawn through a point not on a given line is
   a. greater than 1.
   b. infinite.
   c. 1.
   d. 0.

7. In triangle *ABC*, tangent *A* equals
   a. a/c.
   b. a/b.
   c. b/a.
   d. b/c.

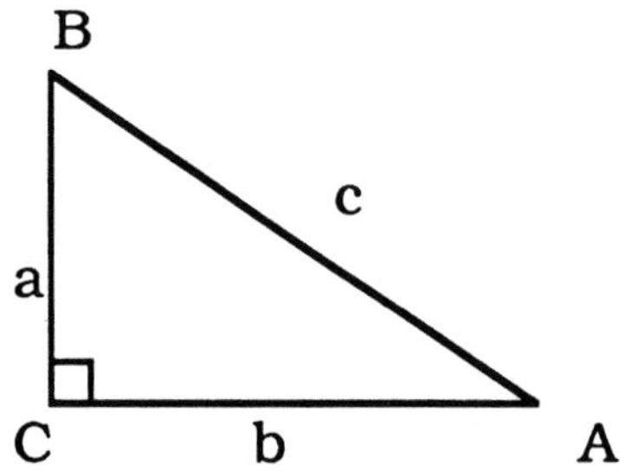

8. The measurement of the Earth's size by Eratosthenes required the use of
   a. the tangent ratio.
   b. non-Euclidean geometry.
   c. Euclidean geometry.
   d. the Pythagorean theorem.

9. The development of non-Euclidean geometry was spurred on by questions raised by
   a. Aristarchus.
   b. Euclid himself.
   c. the right angle postulate.
   d. the parallel postulate.

10. In hyperbolic geometry, the sum of the angles of any triangle is
   a. less than 180°.
   b. equal to 180°.
   c. more than 180°.
   d. more than 180° and less than 360°.

ANSWERS

1. (c), 2. (c), 3. (d), 4. (d), 5. (c), 6. (d), 7. (b), 8. (c), 9. (d), 10. (a).

Sample Problems

11. Illustrate how a spherical triangle can have angle sum equal to 270°. Use a picture and words.

12. Explain why $\sin A < 1$ in any right triangle, ΔABC, where A is an acute angle.

ANSWERS

11.

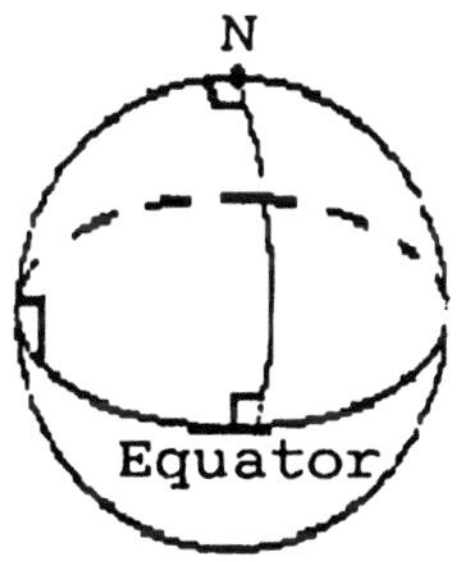

12.

Sin $A = a/c$

Since $c^2 = a^2 + b^2$,

$c > a$. Thus $a/c < 1$.

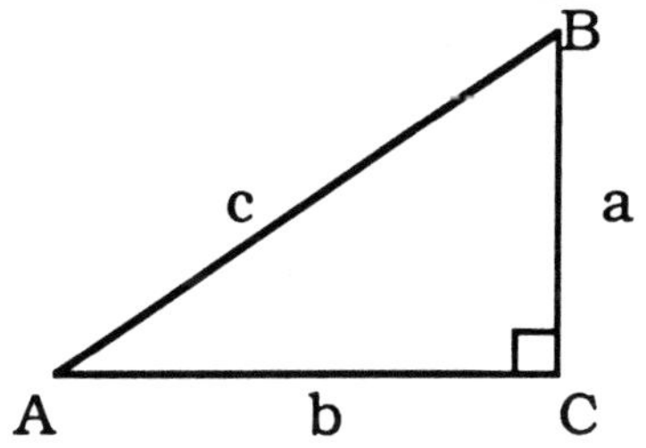

13.

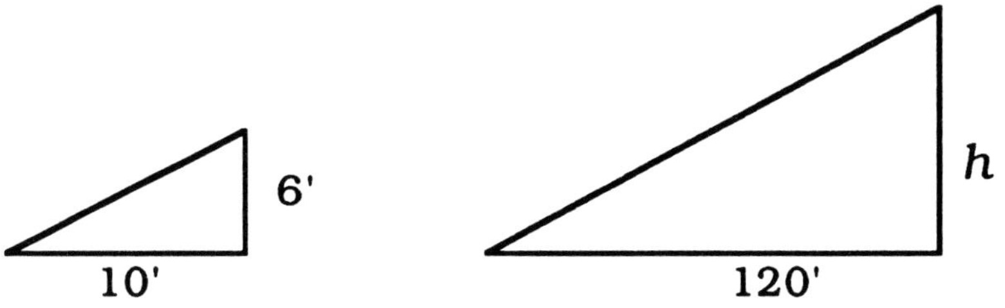

The triangles are similar. Thus $\frac{h}{120'} = \frac{6}{10'}$.

Hence $\frac{6}{10} \times 120' = 72'$.

By the pythagorean theorem,

$Z^2 = (72')^2 + (120')^2 = \sqrt{19584\text{ ft}^2}$

$\therefore Z = \sqrt{19584\text{ ft}^2} = 140\text{ ft.}$

14.

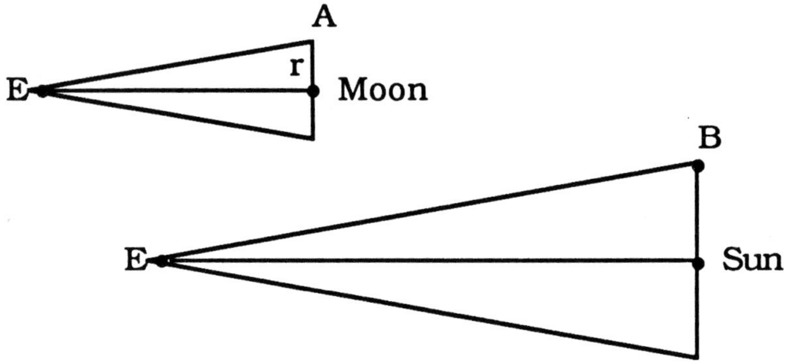

If the disks have the same apparent size, then the triangles ΔEMA and ΔESB are similar.

Thus $\frac{r}{240{,}000\text{ mi}} = \frac{436{,}000\text{ mi}}{93{,}000{,}000\text{ mi}}$, and hence $r = 1125$ mi.

Part V Computer Science
Zaven Karian, Denison University
Wayne Carlson, Cranston/Csuri Productions, Inc.
Sartaj Sahni, University of Minnesota
Paul Wang, Kent State University

INTRODUCTION

In this part, we examine the role of mathematics in the development and workings of modern computational machines. To many people computers and mathematics appear to be one subject, but the relationship is more subtle than that. The roots of the modern computer belong both to early computational devices such as the abacus and to mental constructs which mirror human activity in a search to define mathematical truth.

Improvements and innovations have come at an incredible pace. New developments in computer graphics have led to new mathematical discoveries. Computer verifications have even been incorporated into proofs of major new results. So, in a sense, we've come full circule. The computer began as an idea to help us understand the meaning of mathematical proof. But mathematics is always growing and expanding. Today, the computer not only helps us to do our calculations and draw our pictures, but it is even changing our notion of proof and the very image of mathematics itself.

OVERVIEW SHOW

Computers, as instruments of calculations, have contributed to the development of mathematics. However, the role of mathematics in computer science and its contributions to the development of computers have been more subtle.

The roots of the relationship between mathematics and modern computing go back to the turn of the century when David Hilbert proposed to put all of mathematics on a firm logical foundation. Specifically, Hilbert wanted to show that mathematics was *consistent* (within the framework of an axiomatic system, it is impossible to prove a statement and its negation) and *complete* (everything that is true can be proved). The problem of consistency and completeness became major challenges for mathematicians during the first half of this century.

Traditional mathematicians, who had expected that all of mathematics would be shown to be complete and consistent were disappointed when the Austrian mathematician Kurt Gödel showed that if a mathematical system rich enough to include arithmetic was consistent, then its consistency could not be proved. Furthermore, such systems would necessarily have within them statements that are true but cannot be proved. Gödel's results posed a new problem for mathematicians who now had to be concerned with the truth of a statement and with the provability of statements. To resolve the latter problem, the British mathematician Alan Turing set out to devise a procedure that would determine if a given statement is provable. To put such a procedure to the test, he developed the concept of a machine that would determine the provability of a statement. This conceptual machine was in fact an abstract computer, later designated as a Turing machine and recognized as the theoretical basis of modern computers.

What Turing showed was that a procedure for provability existed if and only if a Turing machine could be implemented to carry out the procedure. From this he was able to conclude that it was impossible to devise a procedure for determining the provability of statements.

SHOW ONE: RULES OF THE GAME: ALGORITHMS

A *procedure* for solving a problem is a sequence of steps that lead to the solution of the problem. For a procedure to be an *algorithm*,

1. each step within the procedure must be *unambiguous*,
2. each step must be *effective* (i.e., feasible),
3. the procedure must produce some output,
4. regardless of the input (if any) the procedure must terminate in a finite number of steps.

In actual practice, algorithms are translated into precisely defined programming languages such as BASIC or Pascal, which in turn are interpreted and executed by a computer. It must be kept in mind that an algorithm is *not* the solution of a problem, but a method through which a solution can be attained. Thus, if an algorithm for solving a problem is available, an understanding of the problem is not essential to finding its solution.

Often a variety of algorithms can be used to solve a given problem. When a choice of algorithms is available, people usually look for a good, or perhaps the best, algorithm. Generally, the terms "good" of "best," when applied to algorithms do not have a precise meaning unless a criterion for making comparative judgements is specified. However, in many situations better means faster.

The program compares two algorithms, the insertion sort and merge sort algorithms, for *sorting* (arranging in a predefined order) objects. The conclusion is that for a large number of objects the merge sort significantly outperforms the insertion sort, even when the merge sort is handicapped by running on a very slow computer. The more detailed comparative analysis done in the text shows that the execution times for sorting n objects via the insertion sort is proportional to n^2 while that of the merge sort is $n \log n$. The text explores in greater detail the implications of various execution times.

The program illustrates two of the many algorithms available for sorting. However, not all problems have algorithmic solutions; in some instances, it is known that a problem does not have an algorithmic solution and in other cases it is not known whether an algorithmic solution is possible.

Skill Objectives
1. To be able to understand the distinction between the solution of a problem and the algorithm which leads to a solution.
2. To be able to follow a simple algorithm and obtain a solution.
3. To be able to give an algorithm for a simple problem.

Self-test

In **Problems 1-8** find the output of the given algorithms. The notation A ← B means assign the values of B to A.

1.
```
1. X ← 1
2. PRINT X
3. X ← X + 2
4. PRINT X
5. X ← X + 2
6. PRINT X
7. END
```

2.
```
1. X ← 1
2. WHILE X < 6 REPEAT
3.        [PRINT X
4.              X ← X + 2]
5. END
```

3.
```
1. X ← 1
2. WHILE X < 100 REPEAT
3.        [PRINT X
4.              X ← X + 2]
5. END
```

4.
```
1. X ← 1
2. WHILE X < 100 REPEAT
3.        [X ← X + 2]
4. PRINT X
5. END
```

5.
```
1. NF ← 1
2. J ← 2
3. WHILE J < 4 REPEAT
4.        [NF ← NF * J
5.              J ← J + 1]
6. PRINT NF
7. END
```

6.
```
1. N ← 10
2. NF ← 1
3. J ← 2
4. WHILE J < N + 1 REPEAT
5.        [NF ← NF * J]
```

```
6.           J ← J + 1]
7. PRINT NF
8. END
```

7.
```
1. S ← 0
2. J ← 1
3. WHILE J < 4 REPEAT
4.       [S ← S + J
5.            J ← J + 1]
6. PRINT S
7. END
```

8.
```
1. N ← 10
2. S ← 0
3. J ← 1
4. WHILE J < N + 1 REPEAT
5.       [S ← S + J
6.            J ← J + 1]
7. PRINT S
8. END
```

ANSWERS

1. 1, 3, 5, 2. 1, 3, 5, 3. 1, ..., 99 all odd integers from 1 to 99, 4. 99, 5. 6, 6. 10! = 3,628,800, 7. 6, 8. 55.

Sample Problems

9. Give an algorithm for finding and printing the sum of the first 4 odd positive integers (i.e., 1 + 3 + 5 + 7)

10. Generalize the algorithm of **Problem 1** to find the sum of the first 100 odd integers. (Hint: use a WHILE construct)

11. Each step in **Problem 1** is executed once. Steps 1 and 5 in **Problem 2** are executed once, and steps, 2, 3, and 4 are executed 3 times each. How many times are each of the steps in **Problems 5** and **6** executed?

ANSWERS

9.
```
1. SUM ← 1
2. SUM ← SUM + 3
3. SUM ← SUM + 5
4. SUM ← SUM + 7
5. PRINT SUM
6. END
```

10.
```
1. SUM ← 0
2. J ← 1
3. NUM ← 1
4. WHILE J < 100 REPEAT
5.       [SUM ← SUM + NUM
6.           NUM ← NUM + 2
7.               J ← J + 1]
8. PRINT SUM
9. END
```

11. For **Problem 5,** steps 1, 2, 6, and 7 are executed once each and steps 3, 4, and 5 are executed twice each.

    For **Problem 6,** steps 1, 2, 3, 7, and 8 are executed once each and steps 4, 5, and 6 are executed 9 times each.

SHOW TWO: COUNTING BY TWO'S: NUMERICAL REPRESENTATION

Because of the two-state nature of the components of modern computers, it is easy to represent information within computers using two symbols--the symbols "0" and "1." A *binary code* consists of *two symbols*, usually designated by "0" and "1" and *a set of rules for interpreting the symbols*. When the meaning of a symbol is determined by the symbol itself as well as the location of that symbol within a sequence, then the coding scheme is called a *place value system*. Binary place value systems are used to represent such diverse forms of information as musical scores, visual images, and numerical data within computing systems.

In about 1850, the British mathematician George Boole developed a binary code in order to study logical inferences. Representing truth by "1" and falsity by "0," Boole developed *truth-values* for statements A AND B and A OR B from the truth-values of the statements A and B. Logical functions such as AND, OR, and XOR (the exclusive OR) are built into modern computers, and through these functions computers can perform a variety of tasks ranging from numerical calculations to the synthesis of music.

Chapters 19 of the text deals with different aspects of binary coding, including how binary codes can be used to store and manipulate integer data. Portions of Chapter 19 on decoding are more directly related to the material covered by the next program.

Skill Objectives

1. Understand the role of binary codes in storing various types of information in computers.
2. Develop truth-tables for simple logical functions.
3. Write positive integers in binary form.

Self-test

1. Determine the truth value (i.e., designate false by 0 and true by 1) of the statements A, B, and C.
   A: George Boole used binary codes before computers were developed.
   B: Computers use binary codes only for storing numbers and musical scores.
   C: Computers are the only devices which use binary codes.

2. For the statements A, B, and C of **Problem 1**, determine the truth values of
   a. A AND B
   b. A AND C
   c. B AND C
   d. A OR B
   e. A XOR B
   f. B XOR C
   g. A OR (B OR C)

Problems 3-7 are multiple-choice questions.

3. The compound statement A AND B is true if and only if
   a. A is true and B is false.
   b. A and B are both false.
   c. A and B are both true.
   d. exactly one of A and B is true.
   e. at least one of A and B is true.

4. The compound statement A OR B is true if and only if
   a. A is true but B is false.
   b. A and B are both false.
   c. A and B are both true.
   d. exactly one of A and B is true.
   e. at least one of A and B is true.

5. The compound statement A XOR B is true if and only if
   a. A is true but B is false.
   b. A and B are both false.
   c. A and B are both true.
   d. exactly one of A and B is true.
   e. at least one of A and B is true.

6. (A AND B) OR C is true whenever
   a. A is true.
   b. B is true.
   c. C is true.
   d. A is false.

7. A AND (B OR C) is false whenever
   a. A is true.
   b. B is true.
   c. C is true.
   d. A is false.

ANSWERS

1. A:1, B:0, C:0, 2. a:0, b:0, c:0, d:1, e:1, f:0, g:1, 3. c, 4. e, 5. d, 6. c, 7. d.

Sample Problems

8. If a musical instrument had 2 notes, a "0" and a "1" could be assigned to each note so that the two notes could be represented by a single bit. How many bits would be required if the instrument had 4, 8, 16, 32, or 256 notes? A standard piano has 88 notes; how many bits would be needed to represent the notes of the piano?

9. A mathematical operation, designated by "*," is called associative if $a*(b*c) = (a*b)*c$ for all possible operands *a*, *b*, and *c*. Use a truth table to determine if the AND logical operation is associative. Describe in words the circumstance under which A AND B AND C will be true.

ABC	A AND B	B AND C	A AND (B AND C)	(A AND B) AND C
FFF	F	F	F	F
FFT	F	F	F	F
FTF	F	F	F	F
FTT	F	T	F	F
TFF	F	F	F	F
TFT	F	F	F	F
TTF	T	F	F	F
TTT	T	T	T	T

10. An operation "*" is distributive with respect to another operation "@" if A*(B@C) = (A*B)@(A*C). Determine if the logical AND operation is distributive with respect to the OR operation.

ABC	A AND B	A AND C	(A AND B) OR A AND C)	B OR C	A AND (B OR C)
FFF	F	F	F	F	F
FFT	F	F	F	T	F
FTF	F	F	F	T	F
FTT	F	F	F	T	F
TFF	F	F	F	F	F
TFT	F	T	T	T	T
TTF	T	F	T	T	T
TTT	T	T	T	T	T

ANSWERS

8. 2, 3, 4, 5, and 8 bits would be required for 4, 8, 16, 32, and 256 notes respectively. For the piano, with 6 bits $2^6 = 64$ notes can be coded; with 7 bits up to $2^7 = 128$ notes can be coded. Therefore 7 bits will be required.

9. A AND B AND C will be true if and only if A, B, and C are all true.

10. From the above truth table we see that values of (A AND B) OR (A AND C) coincide with those of A AND (B OR C) for all combinations of A, B, and C. Thus, the AND operation is distributive with respect to the OR operation.

SHOW THREE: CREATING A CODE: ENCODING INFORMATION

Since computers are constructed from two-state electromagnetic components, it is natural to use a *binary coding* scheme to represent information inside a computer. Any information that can be coded, can be coded in binary form. Furthermore, an unlimited number of coding schemes can be used to represent a given type of data.

Economy of the number of bits used for storing a data item is one of the important criteria upon which the choice of a particular code is based. One bit can represent 2 objects (coded 0 and 1), two bits can represent 4 objects (coded as 00, 01, 10 and 11), and generally k bits can represent $2k$ objects.

Textual data composed of sequences of alphabetic, numeric, and special (punctuation marks, commas, etc.) characters often need to be stored in computers. ASCII (American Standard Code for Information Interchange) is a commonly used coding method for storing textual data. This coding scheme, discussed in considerable detail in Chapter 19, uses 8 bits per character.

Special mathematical techniques can be applied to reduce the number of bits required to store other forms of data such as visual images. Through the use of Laplacean pyramid, a picture can be represented by about 65,000 bits instead of 500,000 bits.

In special applications where the presence of errors is a major concern, codes that can *detect* errors, and possible even *correct* errors,are used. The *Hamming code* is a single error detecting and correcting code that uses 3 *redundant* bits in order to store 4 data bits. The coding mechanism and the detection and correction of errors are discussed more fully in Chapter 19. More complicated codes are used when it is necessary to detect and correct a large number of errors. Such codes are used in conjunction with such binary storage devices as compact discs.

Skill Objectives

1. Understand that a binary code (in fact many binary codes) can be devised to store a given type of data.
2. Determine the number of bits required for the storage of certain types of data.
3. Trace the error detection/correction mechanism embedded in the Hamming code.

Self-test

1. What is the decimal value of the binary number 11...1 (n consecutive 1's)? Hint: add 1 to 11...1.

2. If 4 bits are used to store positive integers in a computer, what is the largest positive integer that can be stored?

3. If n bits are used to store positive integers in a computer, what is the largest positive integer that can be stored?

4. In an 8-bit two's complement representation of integers, what integer is represented by 0100 0001?

5. Suppose B has truth value 1.What is the result of A XOR B when A has value 1? What if A has value 0?

6. If B has truth value 1, describe the value of A XOR B in terms of the value of A.

7. If B has truth value 0, describe the value of A XOR B in terms of the value of A.

8. Illustrate how four bit patterns, 0011 is stored using the Hamming code.

ANSWERS

1. $2^n - 1$, 2. 15, 3. $2^n - 1$, 4. 65, 5. 0 when A has value 1, 1 when it has value 0, 6. A XOR B will be the opposite of A, 7. A XOR B will be the same as A,
8.

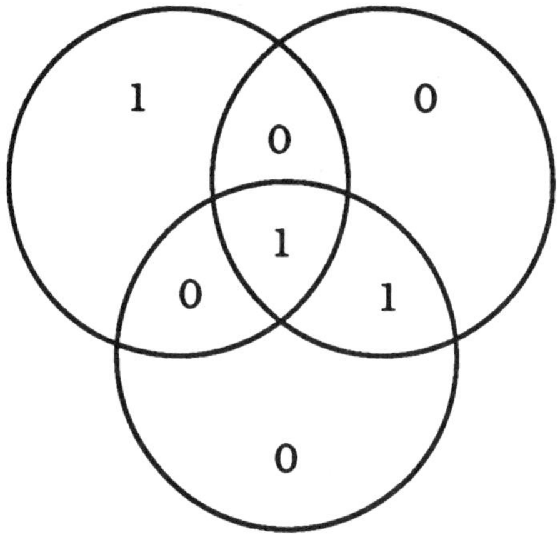

Sample Problems

9. The HBO encryption method uses the XOR operation to combine the bits from a random bit generator with the bits of a password. Later, the same random bits are combined through XOR with the encrypted password to produce the original password. This suggests the following identity

   (P XOR R) XOR R = P

   Use truth tables to verify this identity.

10. ASCII sequentially assigns codes of 0100 0001 (decimal 65) up to 0101 1010 (decimal 90) to the letters "A" through "Z." Lower case letters are sequentially assigned codes of 0110 0001 (decimal 97) for "a" through 0111 1010 (decimal 122) for "z." Numerically, the code for a lower case letter can be obtained by adding 0010 0000 (decimal 32) to the code of the corresponding capital letter. Find an 8-bit pattern, X, such that if Y is the ASCII code for a letter (upper or lower case) then Z = X XOR Y will produce the ASCII code of the same letter with the case reversed.

ANSWERS

9.

P	R	P XOR R	(P XOR R) XOR R
0	0	0	0
0	1	1	0
1	0	1	1
1	1	0	1

10. Take X = 0010 0000. As indicated in **Problems 6 and 7**, when X is XORed with an 8-bit code Y, Z = X XOR Y will have the same bits as X except for the 6th bit (from the right) which will be reversed. Reversing the 6th bit of an ASCII code has the same effect as adding (or subtracting) 32 from its decimal value.

SHOW FOUR: MOVING PICTURE SHOW: COMPUTER GRAPHICS

Pixels are the smallest elements that can be displayed on a computer graphics screen. Similar to graph paper, the screen has rectangular coordinates, and figures can be produced by turning selected pixels "on."

The mathematical formulation of geometric objects (e.g., lines, polygons, etc.) is *idealized* in that such objects consist of infinitely many points. Since a computer rendering of figures must of necessity be composed of a finite number of pixels, jagged edges can result. One common technique for anti-aliasing (reducing the jaggedness) is through shading or coloring pixels along edges according to the proportion of the pixel area within the figure.

Coloring a polygon can be done through several methods. The one described in the program forms triangles with a fixed external point and two successive vertices of the polygon and reverses (foreground to background and background to foreground) colors within the triangle. This is repeated successively with all adjacent pairs of vertices of the polygon. Chapter 20 gives an alternate method for colorizing a polygon.

There are a number of methods for producing a smooth curve in the proximity of a set of points on the screen. The method of *splines* is shown in the program. Fitting quadratic polynomials through sets of 3 consecutive points is another commonly used technique.

Animation, or motion, can be produced by moving and/or restructuring objects that are already on the screen. The basic geometric *transformations* that are used are *translation* (the physical displacement of an object without altering its shape of orientation), *scaling* (changing the size of an object through sketching or shrinking), *rotation* (moving the points of an object counterclockwise about a given reference point), and *reflection* (moving an object to its mirror image across an axis of reference).

For the generation of surfaces, a *light source* is used to simulate different levels of brightness in order to produce a three-dimensional effect. This is done by dividing the surface into small regions and computing a *normal line* (the line perpendicular to the surface) in each region. The more parallel a normal line is to the line from the light source, the brighter the rendering of that region is made.

Skill Objectives

1. Understand the discrete nature of a graphics screen composed of a finite number of points.
2. Trace the coloring of a polygon via the procedure described in the program.
3. Hand-sketch one or two iterations for the construction of a spline curve through a few points.

Self-test

1. Suppose a translation moves the triangle with vertices (0, 0), (1, 0) and (1, 2) 2 units to the right and 3 units up. What will be the coordinates of the vertices of the triangle in its low position?

2. Consider the reflection of the original triangle given in **Problem 1** in the x-axis. What will be the coordinates of the reflected triangle be?

3. Do **Problem 2** with a reflection in the y-axis.

4. If the rectangle with vertices (0, 0), (1, 0), (1, 2) and (0, 2) is rotated (counterclockwise) by an angle of 90°, what will be the vertices of the resulting rectangle?

5. If the original rectangle given in **Problem 4** is stretched by a factor of 2 in the vertical direction and shrunk by a factor of ½ in the horizontal direction, what will be the vertices of the resulting rectangle?

ANSWERS
1. (2, 3), (3, 3) and (3, 5)
2. (0, 0), (1, 0) and (,1 -2)
3. (0, 0), (-1, 0) and (-1, 2)
4. (0, 0) (0, 1), (-2, 1) and (-2, 0)
5. (0, 0), (0, ½), (½, 4) and (0, 4)

Sample Problems

6. Connect the points (1, 1), (10, 10), (17, 1) and (28, 10) by straight lines and draw the "curve" that would result after one iteration in the construction of the spline curve.

7. Show the result of **Problem 6** after a second iteration.

8. Suppose a four sided polygon *ABCD* has the same blue color as its background. To color the polygon red, take a point, P, outside the polygon and reverse the red and blue colors in the triangles *PAB, PBC, PCD,* and *PAD* in succession and verify that at the end of this process, *ABCD* will be red on a blue background.

ANSWERS

6.

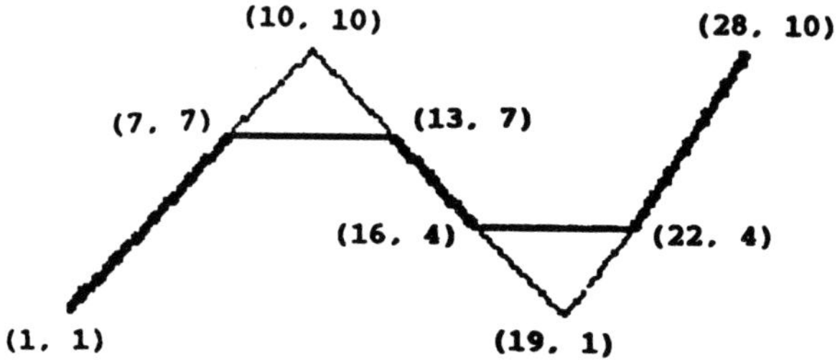
(10, 10)
(28, 10)
(7, 7)
(13, 7)
(16, 4)
(22, 4)
(1, 1)
(19, 1)

7.

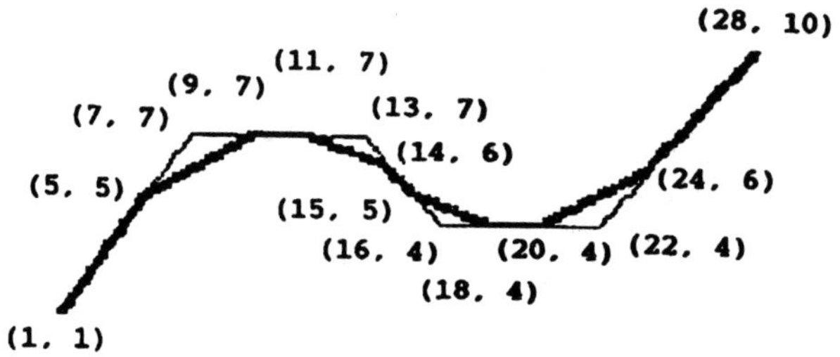
(28, 10)
(11, 7)
(9, 7)
(7, 7)
(13, 7)
(14, 6)
(5, 5)
(24, 6)
(15, 5)
(16, 4)
(20, 4)
(22, 4)
(18, 4)
(1, 1)